BIOENGINEERING FOR POLLUTION PREVENTION

Environmental Research Advances Series

Environmental Research Advances. Volume 1
Harold J. Benson
2008 ISBN 978-1-60456-314-6

Bioengineering for Pollution Prevention
John G. Taylor
2009 ISBN: 978-1-60692-900-1

BIOENGINEERING FOR POLLUTION PREVENTION

DIANNE AHMANN
AND
JOHN R. DORGAN

Nova Science Publishers, Inc.
New York

LIBRARY OF CONGRESS CATALOGING-IN-PUBLICATION DATA
Available upon request

ISBN: 978-1-60692-900-1

Published by Nova Science Publishers, Inc. ✚ New York

CONTENTS

PREFACE

This book looks at petroleum-based fuels and related materials that are central to the economies of developed and developing countries around the world. However, these resources are finite and expected to enter a period of diminishing availability within the next several decades. To move economies based on petroleum and its feedstocks to fuels and materials that are renewable, environmentally friendly, and of greater availability, the science and engineering communities worldwide are exploring many options. As discussed in this book, principal alternative energy resources that scientists have been exploring are wind, solar radiation, hydropower, geothermal power, coal combined with carbon sequestration, hydrogen, and biomass. In addition, biomass and biologically-generated polymers are attractive renewable feedstocks as energy-producing materials. It appears likely that no single resource will offer the versatility of petroleum in the future. As a result, several complementary technologies are being explored to meet the world's diverse needs for energy and resource materials.

Excerpted from State of the Science Report EPA/600/R-07/028, by: Dianne Ahmann and John R. Dorgan.

DISCLAIMER

The information described in this document does not necessarily reflect the views of the Agency, and no official endorsement should be inferred. Mention of trade names or commercial products does not constitute endorsement or recommendation by EPA for use.

ACKNOWLEDGMENTS

This document was reviewed by both internal U.S. Environmental Protection Agency (EPA) reviewers and external peer reviewers who were chosen for their diverse perspectives and technical expertise in bioengineering and pollution prevention. Richard Engler and Mark Segal from EPA's Office of Prevention, Pesticides, and Toxic Substances and Bob Frederick and Diana Bauer from the EPA's Office of Research and Development provided comments on the initial draft. Four external peer reviewers reviewed the final version including: David B. Levin, University of Victoria; Anastasios Melis, University of California, Berkeley; Lonnie Ingram, University of Florida; and Richard Wool, University of Delaware.

EXECUTIVE SUMMARY

Petroleum-based fuels and related materials are central to the economies of developed and developing countries around the world. However, these resources are finite and expected to enter a period of diminishing availability within the next several decades.

To move economies based on petroleum and its feedstocks to fuels and materials that are renewable, environmentally friendly, and of greater availability, the science and engineering communities worldwide are exploring many options. Principal alternative energy resources that scientists have been exploring are wind, solar radiation, hydropower, geothermal power, coal combined with carbon sequestration, hydrogen, and biomass. In addition, biomass and biologically-generated polymers are attractive renewable feedstocks as energy-producing materials. It appears likely that no single resource will offer the versatility of petroleum in the future. As a result, several complementary technologies are being explored to meet the world's diverse needs for energy and resource materials.

Biologically based transformations have several potentially favorable attributes. They typically operate on renewable resources, at low temperatures, in aqueous environments, and produce few byproducts because of the specific nature of enzymatic catalysis. These attributes make industrial biotechnology inherently consistent with the principles of Green Chemistry and promise industrial commodity production with less environmental impact.

In this document, the application of industrial biotechnology to the important commodity classes of fuels and plastics is reviewed. Where applicable, those areas that have been advanced under funding from the joint EPA and National Science Foundation (NSF) program, Technology for a Sustainable Environment (TSE), are highlighted. Promising areas for future exploration and development are identified as well.

A. BIOMATERIALS

The worldwide production of plastics reached 260 billion pounds per year at the end of the 20th century, with a value of over \$310 billion to the U.S. economy in 2002. Approximately one-third of all plastics produced are intended as disposable packaging, and nearly all these plastics are derived from petroleum and are highly resistant to natural biodegradation. Because plastic is recycled at rates of only a few percent in most countries, plastics are rapidly accumulating in unproductive and virtually permanent landfills. Additionally, pollution results from the manufacture, use, and disposal of plastic materials. Notwithstanding, plastics have significant benefits for society, such as abundant sterile medical supplies; increased agricultural production; reduced food spoilage; reduced fuel consumption in lighter-weight vehicles; and, low-cost, net-shape manufacturing. Increases in oil prices that consumers are experiencing at the gas pump are also impacting the plastics industries where production costs are rising and being passed on to the consumer. While energy recovery through combustion, recycling, and minimizing plastics use can all aid pollution prevention, the new industrial biotechnology paradigm can be a more environmentally benign solution to the societal, economic, and environmental impacts of increasing oil consumption

Economically competitive properties are now well within reach among biomaterials including starch cellulosics, proteins, polylactides, soy and plant oil-based plastics, and polyhydroxyalkanoates.

Several research priorities have been identified to improve production of biomaterials. The primary need is the development of standards for assessing the energy content and emissions profiles for plastic materials. This is essential to ensure the pursuit of truly environmentally benign materials.

Metabolic engineering routes to the synthesis of monomers and polymers should also be explored due to the increasing practicality of inducing the necessary genetic changes. Advances in metabolic pathway modeling can further enhance these efforts by indicating promising targets for genetic engineering. Among these, modifications necessary to enable use of waste biomass feedstocks, such as lignocellulosics and waste oils, or to enable biosynthesis within crops that can be grown on marginally productive lands, such as switchgrass, would be especially valuable.

Biopolymers and other bioplastics could be more widely used in place of petroleum-based plastics if their physical properties (such as heat resistance and moisture permeability) could be improved. Conventional composites and nanocomposites in which the reinforcing agents are also based on renewable resources (biocomposites) are of particular interest in this context. In addition, biopolymers based on inexpensive monomers presently available using conventional fermentation technologies should be explored and developed where feasible.

B. BIOFUELS

Energy is a central issue in economic sustainability. Production and distribution of inexpensive energy in a variety of forms (electricity, heating and transportation fuels) is essential for maintaining industries and for supporting stable lifestyles for people. Transportation fuels dominate use of imported petroleum.

In the area of biofuels, fostering collaboration of scientists and engineers is critical. The absence of interdisciplinary collaboration is repeatedly cited as one of the greatest limits to the scale-up and commercialization of bioenergy technologies. Collaborations between those attempting to understand and engineer the organisms and those attempting to design optimal bioreactors and bioseparation processes should be encouraged and when appropriate, should include researchers from industrial laboratories. Such teams are essential to commercialization of the advances made in academic and governmental laboratories.

Within the field of bioethanol production, the most important challenge is the development of feedstocks based on waste biomass. Research in this area will end or reduce reliance on crops produced with conventional energy-intensive practices that makes bioethanol no more sustainable than fossil fuels. Obstacles to waste-based bioethanol include the absence of high-performance, low-cost cellulase enzymes and/or cellulolytic organisms; the separation of lignin from cellulose; the optimization of simultaneous fermentation of hexoses and pentoses; and the purification of the ethanol and recovery of other valuable byproducts.

Biodiesel development is also dependent on non-sustainable agriculture. Therefore, sustainable production of oilseed crops and the development of technologies to allow use of waste oils are important priorities. Following these, bioengineering advances are needed to be efficient in transesterification and separations.

Relative to other biofuels, biohydrogen is still in its infancy. Biohydrogen technology is actually five distinct technologies (direct photolysis, indirect photolysis, photofermentation, water-gas shift production, and dark fermentation) involving four very different types of microorganisms (green algae, cyanobacteria, purple non-sulfur bacteria, and anaerobic heterotrophic bacteria, respectively). While each type of technology faces its own particular challenges, a few priorities are common to all.

Of utmost importance to the three photolytic technologies is the reduction of photosynthetic antenna pigments. Second, the addition of heterologous pigments to allow utilization of photons outside the photosynthetic spectrum is also desirable. Vital to the two non- photolytic technologies is the development of sustainable organic carbon sources. Current efforts underway to use various waste sources as substrates should be strongly encouraged to continue.

Within the realm of biorefinery platform technologies, the development of integrated bioreactor-bioseparation unit operations should be supported, especially those that can overcome inherent limitations of bioprocessing and other integrated designs that selectively remove the components limiting organism growth. Integration offers tremendous benefit for relatively little investment. Bioreactor design and operation can be further optimized with the assistance of improved theoretical models for reaction kinetics, including structured models. This would provide commercial viability for commodity products with narrow profit margins and for membrane-based separation technologies. In addition, new benign solvent extraction processes are needed for bioseparations to avoid fossil-based and/or toxic solvents—supercritical CO_2 is an excellent example. The development of new strategies to suppress or control membrane fouling in relevant separations would also be immensely useful.

C. RESEARCH STRATEGIES

Fostering interdisciplinary research can be accomplished easily by encouraging multidisciplinary teams through the research grant award process. Also, certain elements of the needed bioengineering platform technologies are already well supported by the USDA and to a lesser extent by the DOE. A joint program between the USDA/DOE that supports the development of bioengineering for energy production is in place. Additional partnerships with these agencies would help to improve the knowledge base of bioengineering for pollution prevention.

LIST OF ACRONYMS AND ABBREVIATIONS

^{13}C	Carbon 13
^{14}C	Carbon 14
2D	2-Dimensional
3D	3-Dimensional
3G	1,3 -Propanediol
Al	Aluminum
ANNs	Artificial Neural Networks
AOX	Alcohol Oxidase
ATP	Adenosine Triphosphate
ASTM	American Society for Testing and Materials
B2020	percent Biodiesel combined with 80 percent Petroleum Diesel
B100	100 percent Pure Biodiesel
BCI	BC International Corporation
BRI	BioEngineering Resources, Inc.
BSPs	Biomass Support Particles
CA	Cellulose Acetate
CAB	Cellulose Acetate Butyrate
CAFI	Consortium for Applied Fundamentals and Innovation
CAP	Cellulose Acetate Propionate
CBP	Consolidated Bioprocessing
C–C	Carbon-carbon Bond
cDNA	complementary DNA
CDP	Cargill with Dow Polymers
CFD	Computational Fluid Dynamics
CH4	Methane
CIE	Compression-injection Engines
CO	Carbon Monoxide
CO2	Carbon Dioxide
CODH	Carbon Monoxide Dehydrogenase
CUBIC	Columbia University Bioinformatics Center
DBT	Dibenzothiophene
DNA	Deoxyribonucleic Acid
DOE	Department of Energy
DSA	Diafiltration Saccharification Assay

dsRNA	double-stranded Ribonucleic Acid
Dsz	Biodesulfurization
DXO	1 ,5-Dioxepan-2-one
EMBL	European Molecular Biology Laboratory
EPA	Environmental Protection Agency
Fd	Ferredioxin
FFV	Flexible-fuel Vehicles
GCN	General Control
H+	Hydrogen Ion
H2	Molecular Hydrogen
HA	Hydroxyalkonate
IPTG	Isopropyl-β-D-thiogalactoside
ISO	International Organization for Standardization
JGI	Joint Genome Institute
LCA	Life-cycle Analysis
LCIA	Life-cycle Impact Assessment
LDPE	Low-density Polyethelyne
MF	Microfiltration
MFA	Metabolic Flux Analysis
Mg	Magnesium
MGD,	L-3-methyl glycolide
Mg2+	Magnesium (II) Ion
Mn2+	Manganese (II) Ion
mRNA	messenger RNA
MTBE	Methyl Tertiary Butyl Ether
MW	Molecular Weight
NADPH	Reduced form of Nicotinamide Adenine Dinucleotide Phosphate
NADP	Nicotinamide Adenine Dinucleotide Phosphate
NCBI	National Center for Biotechnology Information
NIH	National Institutes of Health
NMR	Nuclear Magnetic Resonance
NREL	National Renewable Energy Laboratory
NRMRL	National Risk Management Research Laboratory
NSF	National Science Foundation
PCR	Polymerase Chain Reaction
PE	Polyethylene
PEA	Polyester Amide
PET	Poly(ethylene terephthalate)
PHA	Polyhydroxyalkanoate
PHB	Polyhydroxybutyrate
PLAs	Polylactic Acids
Poly (3HB)	Poly [(R)-3-hydroxybutyrate]
PP	Polypropylene
PPM	Primary Packaging Material
ppm	parts per million
PS	Polystyrene

PS I   Photosystem I
PS II   Photosystem II
PVC   Polyvinyl Chloride
[(R)-3HAs]   (R)-3 -hydroxyalkanoates
RISC   RNA-induced Silencing Complex
RNA   Ribonucleic Acid
SEM   Scanning Electron Microscope
siRNA   short interfering RNA
Sn   Tin
SO_2 Sulfur Dioxide
SO_x Sulfur Oxides and Dioxides
SSCF   Simultaneous Saccharification and Co-fermentation
SSF   Simultaneous Saccharification and Fermentation
ssRNA   single-stranded RNA
STRs   Stirred-tank Reactors
SV40   Simian Virus 40
TCA   Tricarboxylic Acid
TEC   Triethyl Citrate
TMC   Trimethylene Carbonate
TPS   Thermoplastic Starch
TSE   Technologies for a Sustainable Environment
UF   Ultrafiltration
USDA   United States Department of Agriculture
UV   Ultraviolet
VOC   Volatile Organic Compounds
WAXS   Wide-angle X-ray Scattering
Zn   Zinc

Chapter I

INTRODUCTION

A. CURRENT STATUS OF ENERGY AND MATERIALS FEEDSTOCKS

1. The Petroleum Resource

Resource shortages are a natural consequence of the utility of the resource combined with human ingenuity. As a new resource is discovered, people invent uses for it in proportion to its adaptability; increased uses typically increase the demand for it often resulting in increased production, which thereby increases the opportunity for new uses to be discovered, and so on. In the case of petroleum, this cycle has progressed to such an extent that developed world economies are dependent on it for heat, food (through agriculture), shelter (through synthesis of construction materials), and transportation.

In the United States, for example, petroleum use has increased steadily from approximately 9.8 million barrels per day in 1960 to ~21 million barrels per day in 2005 [1]. The present rate of demand increase is ~1 .5 percent per year resulting in U.S. demand expected to increase 37 percent over 2004 levels by 2025 [2]. Within this demand, transportation and industry, the latter including plastics and materials production, consume the greatest shares at ~66 percent and ~25 percent, respectively. Comparable increases have been observed in other developed countries as well [3, 4].

If petroleum were plentiful, widely distributed, and environmentally benign, this situation would be no cause for concern. Unfortunately, though, it is not the case.

First, the demand of the developed economies for petroleum is now reaching a level that is comparable to known reserve limits. For example, an estimated 875 billion barrels of oil have been consumed since the dawn of the oil age [5], while 1.7 trillion barrels remain in proven reserves (within oil fields discovered but not yet pumped out), and another estimated 900 billion remain to be discovered [6]. Of the 875 billion consumed, however, greater than 60 percent (550 billion) have been consumed since 1975. With world demand continuing to rise at ~2 percent per year, the world production peak is estimated to occur between 2026 and 2047 [5]. Within the United States, crude oil production is expected to peak in 2010 [2], and the remainder of the petroleum that is easily accessible given both geological and political constraints is expected to peak in that approximate time frame as well [6].

This raises the second issue, that of petroleum accessibility, given both geological and political constraints. Geological factors dictate that not all petroleum is equally accessible.

Fields that may lie in temperate zones within several hundred feet of the surface have understandably been the first to be exploited. This leaves oil beneath oceans, in remote Arctic regions, tightly associated with sands, and/or laden with impurities as an increasing component of that that remains to be exploited. Technological advances have greatly increased the amount of petroleum that can be extracted from the earth, once discovered, but often at high operational and environmental cost Political factors also contribute to petroleum accessibility. While the United States was once the world's greatest petroleum producer, it and many other developed countries are now net importers: the United States now imports approximately 56 percent of its demand, or ~1 1.2 million barrels per day [7, 8]. The United States is not only the greatest consumer of world petroleum resources–demanding ~25 percent of the global total production [4]–but notable to foreign suppliers, the United States is also the greatest importer, with nearly double the net imports of second-ranked Japan [9]. Of the 1.7 trillion barrels of oil in the world's proven reserves, over half of those are located in the Middle East. Petroleum suppliers are therefore becoming increasingly concentrated in regions of the world, particularly the Middle East, that have historically been politically unstable and/or unfriendly to western interests [6].

Finally, petroleum is far from environmentally benign. Petroleum combustion releases carbonaceous gases, principally carbon dioxide (CO_2), carbon monoxide (CO), and methane (CH_4), as well as sulfurous gases such as sulfur dioxide (SO_2). Decades of climate and atmospheric composition data are now confirming the link between increasing concentrations of greenhouse gases such as those emitted by combustion of fossil fuels and increasing global temperatures [10]. In addition, concerns are growing about the volume of discarded wastes that the United States and other countries produce. Global consumption of petroleum-based thermoplastics, the greatest component, now exceeds 100 million tons per year, of which approximately half is discarded within two years of production. Much of the other half, used to generate products with longer lifetimes, is just beginning to enter the waste stream, with the result that plastic waste generation is expected soon to exceed the growth in consumption [11]. This, in turn, is expected to create a considerable demand on landfill space [12].

An accompanying problem is that wealthy countries can export such wastes to poorer countries. Although these practices have been addressed through measures such as the Basel Ban, diminishing waste volumes is the most straightforward solution to exploitation of vulnerable peoples and natural areas [13]. As a result, the impetus for transition from fossil fuels to renewable energy sources and materials feedstocks is resulting as much from environmental considerations as it is from concerns about future conflict over petroleum resources.

2. Benefits of Petroleum Replacement

Numerous technologies are under development for the replacement of petroleum as the primary energy source and materials feedstock in developed countries. Wind, solar, hydroelectric, geothermal, and biomass-derived power will each be called upon to contribute to the post-petroleum economy, and conservation measures are also expected to receive greatly increased attention. Materials derived from biological molecules are also gaining diversity and availability.

To what extent can petroleum be replaced, and by what alternatives, within the next decades? While this potential is debatable, a realistic best-case scenario can be presented given recent projections. In April, 2005, a report by the U.S. Department of Energy (DOE) and the U.S. Department of Agriculture (USDA) estimated that the United States could quite feasibly produce 1 billion dry tons of biomass feedstock (over half of which is waste) per year, enough to displace 30 percent or more of the present U.S. petroleum consumption for fuels and materials by 2030 [14]. When this is combined with projections of solar and wind energy together providing 20 percent of the power demand in the industrialized world by that time, it appears that biological resources could become important contributors to the evolution of a post-petroleum world. Under these projections, CO2 emissions could peak before 2050 and conventional fossil fuel use could be virtually eliminated by 2100 [6].

3. Unique Contributions from Biotechnology

The application of biotechnology to the production of commodities–notably fuels, chemicals, and structural materials–increases the array of options available to supply sustainable resources and preserve the environment. In particular, through the use of biological feedstocks, biotechnology has the potential to minimize greatly the overwhelming dependence developed countries, particularly the United States, have on petroleum and other non-renewable fossil fuels for production of fuels and plastics.

Numerous mechanical, geothermal, and electrical technologies offer valuable contributions to issues of energy independence and pollution prevention [15]. At the same time, several unique advantages are brought by biofuels and bioproducts that will be highlighted in this report.

First, many biotechnologies use the abundant, renewable, and potentially sustainably-produced resource of plant biomass as the primary feedstock for liquid biofuels, biochemicals, and biomaterials. Cellulosic and other biomass is currently available at the commodity scale and is increasingly cost-competitive with petroleum, especially when environmental costs are included, on both energy and mass bases [16]. Indeed, the land resources of the United States are capable of producing a sustainable supply of biomass sufficient to displace 30 percent or more of the country's present petroleum consumption, amounting to approximately 1 billion dry tons of biomass feedstock per year [14].

Second, microbiotechnologies have the potential to use simple, organic, and inorganic feedstocks in microbe-based bioreactors that generate desired products directly, without plant biomass intermediates. For example, photosynthetic microbial biohydrogen production requires only sunlight, CO2, salts, and water [17], and bioplastic precursors such as polylactic acid and polyhydroxybutyrate can be made directly by microbes as well [18, 19].

Third, biotechnologies make use of enzymes, proteinaceous catalysts that are often exquisitely selective and provide high rates of product generation. Unlike other catalysts, enzymes can be manipulated genetically to improve parameters such as substrate affinity, specificity, and catalytic rate, as well as tolerance to process conditions, longevity, and even production rate of the enzyme itself by the host cell.

Finally, as a result of the above, many biotechnologies are able to avoid use of toxic feedstocks and processing reagents that are necessitated by conventional methods and thereby minimize toxic wastes. For example, biosynthesis of the denim dye, indigo, requires only glucose as substrate, in contrast to the conventional synthesis that requires benzene or other aromatic solvents [20].

Bioengineering for pollution prevention is an emerging area of both an intellectual endeavor and an industrial practice. The economic driving forces, the importance of feedstock, and the scale of production all distinguish this arena of biotechnology from the pharmaceutical and nutritional sectors. While fossil fuel-based economies typically evolve from a relatively low-value commodity (e.g., kerosene for lighting) to intermediate-value materials (gasoline, plastics) and ultimately to valuable specialty chemicals (cosmetics, pharmaceuticals), it appears that the biobased economy is progressing from high-value products (pharmaceuticals) to those of intermediate value (industrial catalysts, plastics). As biotechnology evolves and matures, the production of large-scale, relatively low-value products such as fuels is becoming increasingly attractive and economically feasible.

B. CURRENT CHALLENGES

1. Cellulose Stability

The greatest impediment to widespread application of bioengineering for production of commodities is currently the general absence of low-cost processing technologies for biomass. Challenges associated with the conversion of plant biomass into useful products are dominated by the chemical stability of cellulose within biomass (Figure 1), causing it to resist modification and therefore require valuable enzymes or other catalysts, as well as special processing conditions [21].

Advances are therefore greatly needed in both enzymatic and non-enzymatic biomass pretreatment technologies, as well as in the development of efficient product-producing microbes and fermentation bioreactor technologies, the latter of which would directly benefit non-biomass-consuming processes as well.

The generation of high-value coproducts has the potential greatly to offset expenses of processing any feedstock, showing that exploration of the diversity of products that a process or feedstock can yield is also of central importance to the realization of true, profitable, economically resilient biorefineries.

2. Engineering Optimal Organisms

Genetic and metabolic engineering techniques are now being used to address the mivrobial and enzymatic problems of biocommodity production from hundreds of different angles. The central goal, in virtually all cases, is the development of organisms that can use low-cost substrates, give high product yields, and/or exhibit robustness in temperature and pH extremes characteristic of many industrial environments.

The rapidity of development of techniques for manipulation and/or analysis of gene sequences and expression patterns, as as the exponential rate of accumulation of genetic information about numerous industrial microorganisms, are enormous forces propelling the field of bioengineering forward [22].

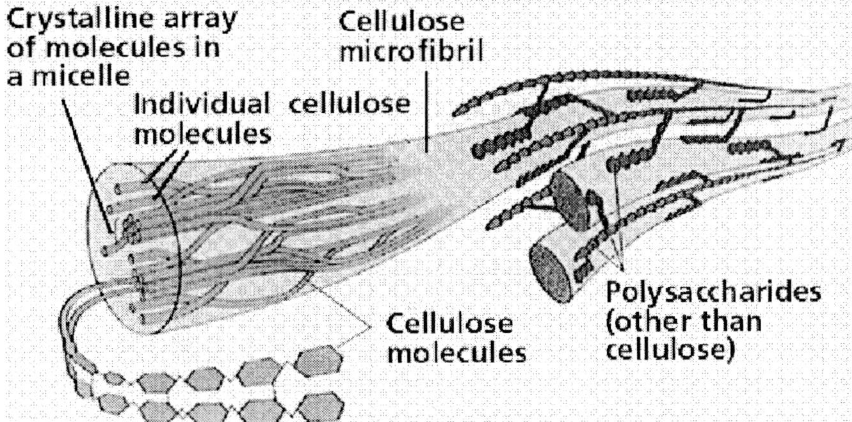

Reprinted from Purves et al., Life: The Science of Biology, 4th Edition, by Sinauer Associates (www.sinauer.com) and WH Freeman (www.whfreeman.com), with permission.

Figure 1. Structure of cellulose as it occurs in a plant cell wall.

3. Bioreactors and Bioseparations

Similarly, reactor and separations technologies must play important roles in the development of economically feasible biorefineries, as these processing steps often contribute the greatest expense of the final products. At early stages of development, new bioengineering processes often benefit greatly from combination of separate components. For example, integrating unit operations, as into reactor-separators and other novel processes, often provides immediate improvements in efficiency. Similarly, integrating individual production steps into a multi-product biorefinery, and integrating biorefineries into the broader economic and environmental systems in which they function, are important avenues to economic feasibility. To evaluate energy, material, and cost efficiencies of these systems, Life-cycle Analysis (LCA) tools are invaluable and are themselves undergoing rapid development [23]. An enormous barrier exists in the funding and deployment of any pioneer manufacturing plant, and federal programs may be necessary to provide loan guarantees and other incentives to encourage enterprise in this direction [24].

Complete sustainability. Important considerations in the overall sustainability of both biofuels and biomaterials include the fossil-energy costs of agriculture and processing. For example, current practices of conventional agriculture use petroleum-powered machinery and fertilizers synthesized with fossil energy. In addition, conventional processing techniques use nonsustainable energy sources. Furthermore, the treatment of acidic, alkali, and/or organic wastes resulting from processing techniques consumes additional energy. As a result, no biofuel or biomaterial is currently completely sustainable as an energy resource or completely free of pollution generation. Bioethanol, in particular, has come under criticism for its use of intensively-cultivated food crops, while several bioplastics potentially use as much fossil energy in their processing as petroplastics use in their feedstocks and processing combined. For this reason, development of technologies that use waste or other biofeedstocks with low-embodied energy (energy consumed in its production) is a high priority for both biofuels and biomaterials as is development of efficient bioprocesses, including enzymatic and therefore low-wastegenerating catalysis. Underlying each of these, moreover, is the importance of developing sustainable agriculture

C. TERMS AND DEFINITIONS

The field of bioengineering for pollution prevention as outlined is immense. A tremendous number of talented and energetic people are working to bring the elements outlined in the preceding paragraphs into reality. To create a coherent, focused, and thorough document, the authors have limited the discussion to the areas of energy and materials, and provide here a set of terms that further define, limit, and clarify the topics involved.

1. Bioengineering

In its broadest sense, this term is applied to the manipulation, influence, or purposeful design of any entity, material, or process involving biological components. For the purpose of this document, however, a more restricted definition is adopted, in which the engineering itself must involve the manipulation and/or exploitation of biological components themselves to accomplish the desired task.

2. Pollution Prevention

In the context of this document, "pollution prevention" is restricted to the description of processes that deliberately avoid the generation of environmentally deleterious substances. Processes that reclaim, recycle, or degrade such substances, once generated, are excluded.

3. Materials

This term is broadly used to represent any form of matter.

4. Biopolymers, Bioplastics

These terms are used specifically to represent high-molecular-weight structural materials that can be shaped or otherwise manufactured into useful articles for human use.

5. Energy and Fuels

"Energy" in this document emphasizes biological, biologically-produced, or biologically-modified materials or processes capable of generating heat or power. Thus, all significant biofuels (bioalcohols, biodiesel, biohydrogen) are included, in addition to conventional fuels that have been enzymatically treated to reduce the generation of pollutants.

6. Economics and Commercialization

In a market economy, the financial aspects of a new technology inevitably have a major impact on its adoption. Indeed, the majority of technological limitations and research needs described below derive their importance from the need to make pollution prevention technologies competitive with their conventional counterparts. Nevertheless, as a review of the state of the science, thorough analyses of the process economics of each technology are beyond the scope of this work. Many excellent reviews of this aspect exist however, and are cited in the text.

D. STATEMENT OF PURPOSE

The purposes of this document are four-fold: first, to explore the realm of current and developing technologies in the fields of biomaterials and biofuels that benefit environmental integrity through their production and use; second, to identify the most promising and most essential areas of endeavor within each field, thus highlighting top priorities for further research and development; third, to examine the technological challenges and/or barriers to the progress of the given technologies; and fourth, to elucidate the contributions of the NSF/EPA's Technology for a Sustainable Environment Program for each topic area.

REFERENCES

[1]   Energy Information Administration (2005). *Annual energy outlook 2005*, http://www.eia.doe.gov/oiaf/aeo/.

[2]   Reuters (2004). *Petroleum demand to grow 37 percent by 2025–EIA*, MSNBC, Washington, D.C.

[3]   U.S. Department of Energy, Energy Information Administration (2003). *Annual energy review 2002*, Washington, D.C., http://www.bts.gov/publications/pocket_guide _to _transportation/2004/html/figure _06 _table.html.

[4]   Bureau of Transportation Statistics (2004). *Overview of U.S. Petroleum Production, Imports, Exports, and Consumption*, Washington, D.C., http://www.bts.gov/publications/ national_transportation_statistics/2004/html/table_04_0 1 .html.

[5]   Wood, J. H., G. R. Long, and D. F. Morehouse (2004). *Long-Term World Oil Supply Scenarios*, Energy Information Agency, http://www.eia.doe.gov/pub/oil_gas/ petroleum/feature articles/2004/worldoilsupply/oilsupply04.html.

[6]   Roberts, P. (2004). *The End of Oil*. Houghton Mifflin, Boston.

[7]   U.S. Department of Energy, Energy Information Administration (2003). *Net Imports of Crude Oil and Petroleum Products into the United States by Country, 2003*, Washington, D.C., Table 29 Petroleum Supply Annual 2003, Volume 1, page 68.

[8]   U.S. Department of Energy, Energy Information Administration (2004). *U.S. Petroleum Imports and Exports*, Washington, D.C., http://www.eia.doe.gov/oil gas/petroleum/ info_glance/importexport.html.

[9]   U.S. Department of Energy, Energy Information Administration (2001). *Top Petroleum Net Importers*, Washington, D.C., http://www.eia.doe.gov/emeu/security/topimp.html.

[10]   U.S. Environmental Protection Agency (2000). *Global Warming–Climate*, http://yosemite.epa.gov/oar/globalwarming.nsf/content/Climate.html.

[11]   Greenpeace (1998). *The Plastics Boom and the Looming PVC Waste Crisis*, Amsterdam, The Netherlands, http://archive.greenpeace.org/comms/pvctoys/reports/ loomingcontents.html.

[12]   Manchester Metropolitan University (2004). *Atmosphere, Climate, & Environment Information Programme. Waste Fact Sheet Series*, Manchester, UK, http://www.ace.mmu.ac.uk/Resources/Fact Sheets/Key Stage 4/Waste/05 .php.

[13]   Basel Action Network (1997). *The Basel Ban. A Triumph Over Business-As-Usual*, Basel, Switzerland, http://www.ban.org/about_basel_ban/jims_article.html.

[14]   U.S. Department of Energy and U.S. Department of Agriculture (2005). *Biomass as Feedstock for a Bioenergy and Bioproducts Industry. The Technical Feasibility of a Billion-Ton Annual Supply*, Oak Ridge, TN, DOE/GO-102995-2135.

[15]   Hester, R. E., and R. M. Harrison, Eds. (2003). *Sustainability and Environmental Impact of Renewable Energy Sources*. Springer-Verlag Telos, New York, NY.

[16]   Lynd, L. R., and C. E. Wyman (1999). *Testimony Before the Senate Committee on Agriculture, Nutrition, and Forestry on Senate Bill S935. Sustainable Fuels and Chemicals Act of 1999*. Thayer School of Engineering, Dartmouth College, Dartmouth, NH.

[17]   Levin, D. B., L. Pitt, and M. Love (2004). *Biohydrogen production. prospects and limitations to practical application*, Intl J Hydrogen Energy 29:173-185.

[18]   Salehizadeh, H., and M. C. van Loosdrecht (2004). *Production of polyhydroxyalkanoates by mixed culture. recent trends and biotechnological importance*, Biotechnol Adv 22:261-279.

[19]   van Maris, A. J. A., W. N. Konings, J. P. van Dijken, and J. T. Pronk (2004). *Microbial export of lactic and 3-hydroxypropanoic acid. Implications for industrial fermentation processes*, Metabo Eng 6:245-255.

[20]   Berry, A., T. C. Dodge, M. Pepsin, and W. Weyler (2002). *Application of metabolic engineering to improve both the production and use of biotech indigo*, J Ind Microbiol Biotechnol 28:127-133.

[21]   Klemm, D., B. Philipp, T. Heinze, U. Heinze, and W. Wagenknecht (1998). *Comprehensive Cellulose Chemistry, Volume 2. Functionalization of Cellulose*. Wiley-VCH.

[22]   Ausubel, F. M., R. Brent, R. E. Kingston, D. D. Moore, J. G. Seidman, J. A. Smith, and K. Struhl, Eds. (1988 [updated quarterly]). *Current Protocols in Molecular Biology*, Ringbound Edition. Greene Publishing Associates.

[23]   Hauschild, M. Z. (2005). *Assessing Environmental Impacts in a Life-Cycle Perspective*, Environ Sci Technol 39:81A-88A.

[24]   Ingram, L. (2005). *Personal communication.*

Chapter II

BIOTECHNOLOGICAL PLATFORMS

A. GENETIC ENGINEERING

1. Introduction

The diversity of biological activity and biological products offered by nature is a great resource for biotechnology; inevitably, however, the limits of natural products or activities are eventually found, and ways are sought to improve them. In these cases, either the environment can be manipulated (as in bioreactors, described separately) or the organism itself can be manipulated physiologically or genetically. The possibilities offered by genetic engineering have grown dramatically in recent decades, and it is currently possible to alter the capabilities of organisms, especially of microbes and plants, such that they exhibit activities and generate products vastly different from those of their genetic precursors. While a thorough review of current genetic engineering technology is beyond the scope of this report, it is nevertheless useful to consider the methods most widely utilized in the development of biofuels and bioplastics. These methods include:

- The cloning of key genes (regions of deoxyribonucleic acid (DNA) that encode enzymes) so that they may be moved and changed at will;
- The detection of gene expression patterns by microarray analysis (facilitating the controlled expression or over-expression of desired genes in native or heterologous hosts);
- The corresponding deletion or repression of expression of undesired genes;
- The use of genomic techniques to predict functions of genetic loci;
- The mutagenesis of genes to provide variants with altered activities, specificities, environmental sensitivities, etc.; and
- The integration of the above methods to construct new biosynthetic pathways.

2. State of the Science

2.1. Cloning and Sequencing

The "cloning" of a gene refers to the physical isolation of a sequence of DNA that encodes a complete polypeptide in a form that can be replicated and manipulated easily.

Typically, this involves extraction of total genomic DNA from an organism, use of restriction enzymes to cut the DNA into manageable fragments, and ligation of the fragments into DNA vectors known as plasmids or cosmids. The collection of fragments is known as a "library," and screening of the library by molecular or genetic methods ultimately leads to identification of the vector bearing the desired gene. Further manipulation, including gene isolation through subcloning of the genomic fragment, is then possible [1].

Gene cloning is a well-established technology that has nevertheless become far more efficient with advances in enzyme and DNA purification technologies. The cloning of industrially-important genes is essential to convenient gene expression because, while desirable genes may be found in numerous organisms, only a few organisms are sufficiently durable and productive for industrial bioprocesses. Because genes are expressed through the action of numerous enzymes that are conserved to some extent among organisms, it is often possible to express a cloned gene from one organism in another; moreover, it is also often possible to control the extent of expression very closely, as described below.

2.2 Genomics

The field of genomics–the study of the sequence, structure, and function of an organism's complete set of genetic information (its genome) [2]–is expanding rapidly due to the invention of numerous high-efficiency and high-throughput technologies capable of handling large amounts of DNA as well as sequence data. A key player in this field, and especially important from the perspective of environmental biotechnology, is the Joint Genome Institute (JGI) (www.jgi.doe.gov). The JGI was established in 1997 to unite the expertise and resources in genome mapping, DNA sequencing, and information sciences among the DOE genome centers at the Lawrence Berkeley National Laboratory, the Lawrence Livermore National Laboratory, and the Los Alamos National Laboratory. Its mission is to advance high-throughput, genomescale, computational technologies that facilitate understanding of relationships among genome structure and function, and it now has the capacity to generate DNA sequences of two billion nucleotide bases per month. The JGI contributed complete sequences of Chromosomes 5, 16, and 19 to the Human Genome Project but now, in contrast to National Institutes of Health (NIH)- funded genome sequencers that continue to concentrate on human targets and applications, the project has turned its efforts toward the broader biosphere. Importantly, a primary goal of the JGI is to make high-quality genome sequence data freely available to the scientific community through its Web site (http://www.jgi.doe.gov/sequencing/seqplans.html). The sequences of numerous bacteria, fungi, trees, green algae, protists, crop plants, fish, and amphibians, either presently available or are planned to be available, will continue to be a tremendous resource for the development of bioplastics and biofuels as well as other areas of environmental biotechnology.

Another highly important component of the current genomics landscape is GenBank®, the annotated NIH genetic collection that holds all publicly available DNA sequences in a searchable database [3]. This set includes virtually all sequences published, because many journals require submission of sequence information to a database prior to publication and because GenBank exchanges data daily with other members of the International Nucleotide Sequence Database Collaboration (the DNA DataBank of Japan and the European Molecular Biology Laboratory). GenBank is accessible through the National Center for Biotechnology Information (NCBI) search engine, which integrates data from the DNA and protein sequence databases with taxonomy, genome mapping, protein structure, and domain information, as well as journal literature (http://www.ncbi.nlm.nih.gov/Genbank/index.html).

The value of complete, annotated, searchable genome sequence data for environmental biotechnology cannot be overstated. One example is the discovery of genes analogous to a gene of interest that may possess characteristics desirable for industrial processes or that may be desirable in combinatorial mutagenesis (see section 2.5.3). Other examples are the location of transposon-induced mutations or other mutations that have conferred desirable (or undesirable) phenotypes on an organism and the elucidation of the function of a DNA sequence that complements a mutant phenotype. Each of these contributes to the understanding of basic biochemical processes as well as to the development of industrially-useful bioprocesses. In addition, the progress of biotechnology is therefore significantly enhanced by progress in genomics [4, 5].

2.3. Expression Control

Gene expression is the process by which a gene's coded information is converted into active ribonucleic acid (RNA) and subsequently polypeptide molecules. Because the timing and extent of expression is crucial for cellular energetic and catalytic efficiency, transcription is controlled carefully in response to internal and external conditions. Upstream and intron-based DNA sequences known as promoters, enhancers, and other regulatory elements, as well as regulatory proteins known as transcription factors, are important factors in transcriptional control. Transcriptional regulatory elements differ in strength, or the rate at which they direct a cell to transcribe a gene into RNA, as well as in inducibility, or the stimuli to which they respond via transcription factors [6].

2.3.1. Expression enhancement: strong and inducible promoters. Close control of expression of genes that direct synthesis of a desired product is advantageous and often essential to the development of a successful bioprocess, such that the maximum amount of substrate is converted to product while maintaining good health of the culture. One of the earliest and still widely-utilized promoters engineered for exquisite control of heterologous gene expression is the tac promoter, also referred to as Ptac. This sequence was designed from a combination of the inducible lac and trp promoters in the bacterium Escherichia coli, forming a hybrid that directs transcription up to 11 times more efficiently than the parental sequences. This operon is repressible by the lac repressor and is inducible by addition of isopropyl-ß-Dthiogalactoside (IPTG) [7]. Numerous other expression systems have since been designed for both research and industrial purposes, prominently including those based on viral T4, T7, simian virus 40 (SV40), adenovirus, baculovirus, and cytomegalovirus promoters, the bacterial lac, araB, and xyl promoters, and yeast alcohol oxidase (AOX) and general control (GCN) 4 promoters [8, 9].

2.3.2. Expression repression: RNA interference. Alternatively, it is sometimes desirable to prevent the expression of a gene altogether, or at least to diminish it significantly. While the ideal approach to the problem of gene silencing is the deletion of the gene in question, this is often a difficult and time-consuming process. An attractive emerging solution in eukaryotic cells is the use of RNA interference, or RNAi, to diminish the expression of particular genes with great specificity. This method takes advantage of an antiviral and possibly regulatory strategy in which double-stranded RNA (dsRNA), an unusual molecule within a cell, is first bound by a nucleolytic enzyme called Dicer. This enzyme fragments the dsRNA into fragments of 21 to 23 nucleotides in length, known as small interfering RNAs (siRNAs), which bind in turn to an RNA-induced silencing complex (RISC). RISC then uses one strand of siRNA to bind to single-stranded RNA (ssRNA) molecules of complementary sequence, typically messenger RNAs (mRNAs), and cleave them. Because such mRNAs become untranslatable once cleaved, their expression is effectively blocked.

Several methods have been developed for delivery of siRNAs to industrial cells: the chemical or enzymatic synthesis of siRNAs followed by transfection into the target cells, which is rapid and well-suited to preliminary experiments but is expensive and achieves only transient interference; the incorporation of siRNA sequences into a DNA plasmid, which also requires transfection but achieves stable interference once the vector is integrated into the host genome; and incorporation of siRNA sequences into viral vectors, which can achieve stable interference and can propagate themselves among target cells but possess potential biohazard risks [10, 11].

2.4. Microarray Analysis

Essential to the control of gene expression in complicated pathways, such as those involving the transport of mRNA and/or polypeptides among eukaryotic organelles for processing, cofactor insertion, and folding, is often the understanding of expression patterns of the multiple genes involved in the pathway. A tremendous advance in these efforts has been realized recently in the form of gene microarray technology.

Microarrays are ordered sets of DNA molecules of known sequence applied, often robotically, in a grid of tiny spots onto a glass slide coated with an organic compound such as aminosilane to enhance DNA binding. Hundreds to thousands of distinct DNA molecules may be present in a single microarray, or gene chip, and are most often prepared in one of two ways: first, by a photolithographic process in which single-stranded oligonucleotides of unique, desired sequence are synthesized directly on the slide, most often employed in industry, or second, by physically attaching DNA fragments–such as polymerase chain reaction (PCR)-amplified genomic library clones–to the solid substrate, used almost exclusively in academic research. While the former allows higher density of features (>280,000 on a 1 .28x1 .28-cm array) and elimination of the need to generate cloned DNA or PCR products, the latter has lower cost and greater flexibility [12].

For analysis of gene expression patterns in eukaryotes, messenger RNA is collected from cells of interest under two (or more) conditions of interest. Each sample is then reverse-transcribed into complementary DNA (cDNA) using nucleotides labeled with contrasting fluorescent dyes. The cyanine dyes Cy3 and Cy5 are popular, although over 70 different dyes are available [13]. In prokaryotes, a collection of mRNA is impractical due to the absence of polyadenylation; therefore, total RNA must be collected and labeled by covalent linkage or by using labeled, but random oligonucleotide primers during reverse transcription.

On printed (non-photolithographic) DNA microarrays, relative transcript abundance is then measured by hybridizing cDNA samples to the microarray simultaneously and determining the fluorescence ratio, revealing binding of homologous sequences, for each spot on the array.

On photolithographic oligonucleotide arrays, in contrast, multiple probes from the same single gene of interest, each with a corresponding mismatch probe that serves as internal control, as well as known amounts of labeled transcripts for genes that serve as internal standards, are hybridized simultaneously to the microarray to enable quantitation of transcript abundance.

Gene expression experiments can also be performed by hybridizing a single labeled mRNA sample to "macroarrays" of DNA elements that are supported on positively charged filters. Specialty arrays can be made and analyzed by this method relatively cheaply, and human, mouse, and microbial macroarrays are commercially available (SigmaGenosys, The Woodlands, TX; Research Genetics, Huntsville, AL; Clontech Laboratories, Palo Alto, CA; Genome Systems, St. Louis, MO). The major disadvantages of this format are reduced sensitivity, limited numbers of elements, and the need for higher concentrations of labeled cDNA [12].

Microarray technology is sufficiently promising for medical and pharmaceutical applications that it is expected to continue to attract strong commercial interest leading to increasing array element density, greater detection sensitivity, and more cost-effective methods. Additional details and technical descriptions are available in recent reviews [14, 15].

2.5. Mutagenesis

The DNA molecule consists of two strands of nucleotide bases held together by hydrogen bonds between complementary bases: adenosine pairs with thymine, and cytosine pairs with guanine. Within the coding region of a gene, each triplet of nucleotides specifies an amino acid, and the strand of amino acids in turn folds, sometimes after processing, into a functional protein. DNA is replicated during each cell division in a growing organism, and although this process is highly accurate, errors can nevertheless occur in which an incorrect nucleotide is inserted into the daughter strand. If the incorrect nucleotide is copied faithfully in subsequent replication, a mutation is generated that may ultimately affect the function of the resulting protein and give rise to a variant organism [16].

Mutations arise in nature either randomly or as consequences of DNA damage by ultraviolet (UV) radiation, chemicals, or other mutagenic agents [16]. Researchers seeking desirable mutations can accelerate the mutagenic process by four primary methods: stimulating replication error-based mutagenesis with chemicals or radiation; generating localized sequence alterations through error-prone PCR; inducing combinatorial mutagenesis through directed evolution; or using structural information about a protein to change specific amino acids, termed "rational design." Each of the latter three is conducted in vitro and must therefore be followed by reintroduction of the mutated sequence into the host organism. If desired mutations show phenotypes that are recessive to the wild-type phenotype, then the wild-type gene must first be disabled or displaced by integration of the new sequence in its place; often, however, the desired mutations have dominant phenotypes (e.g., enabling function under harsh conditions) so that such measures are not necessary. Nevertheless, expression of a mutant sequence can be vulnerable to the transcriptional environment at its locus of integration. In the case that the mutant sequence must be integrated into host DNA to be maintained stably, numerous transfections may be required before satisfactory expression is obtained

2.5.1. Chemical and physical mutagenesis. Chemical and physical agents that damage DNA induce mutations during subsequent DNA replication. DNA-damaging chemicals include alkylating agents such as ethylmethanesulfonate and derivatives of nitrosoguanidine that attach alkyl groups to DNA bases, which promotes mispairing during subsequent replication. They also include intercalating agents such as ethidium bromide that insert themselves between base pairs, which changes the spacing between bases and therefore induces insertion of extra nucleotides during replication. Finally, DNA-damaging chemicals also include bulky adduct- producing agents such as benzo(a)pyrene that attach themselves to DNA bases, which also promotes mispairing during subsequent replication. Physical mutagens include electromagnetic radiation such as gamma rays, x-rays, and UV light, as well as particle radiation such as fast and thermal neutrons as well as alpha and beta particles [16, 17].

The gene encoding the protein or RNA molecule of interest need not be cloned or even identified for chemical or physical mutagenesis to be used effectively; indeed, only an activity of interest is required. Exposure of the experimental organism to the mutagenic conditions is followed by screening or, ideally, selection for improvement of the trait of interest, where selection refers to a process that favors reproduction of organisms showing the improved trait over those without improvement [18]. Improved mutants are typically mated, or back-crossed to parental strains, to minimize the accumulation of potentially deleterious mutations in non-target genes, before successive rounds of mutagenesis are undertaken [19].

Chemical and physical mutagenesis methods are relatively straightforward and inexpensive; however, additional methods have been sought because these methods typically affect only a few nucleotides at a time and therefore result in limited improvements [20].

2.5.2. Error-prone PCR. Error-prone PCR is widely used to generate random mutants. In this approach, the gene of interest has been identified and cloned, and PCR primers are designed for it. During amplification of the gene, however, a number of changes from normal PCR are employed: primer annealing temperatures are lowered to diminish fidelity; nucleotide ratios may be lowered and/or altered from the normal 1:1:1:1 equivalence; a non-proofreading DNA polymerase is used; high levels of the magnesium (II) ion (Mg^{2+}) and often the manganese (II) ion (Mn^{2+}) are used to further diminish replication fidelity; and up to 80 cycles may be conducted [21–23]. Amplified products must then be re-introduced into the experimental organism and expressed to allow screening or selection for improved mutants [21, 22]. While this method is rapid, convenient, and generates large numbers of mutants, it primarily generates "point" mutations, or mutations in which single, isolated nucleotide changes predominate. As a result, it explores only a small sequence space, meaning it is unable, through successive rounds of mutagenesis, to converge upon a globally optimal sequence [20]. Nevertheless, numerous commercial kits are available to facilitate error-prone PCR, and it remains a popular method of mutagenesis (www.stratagene.com/products/display;www.jenabioscience.com/images/0ea5cbe470/PP-102.pdf).

2.5.3. Combinatorial mutagenesis. Combinatorial mutagenesis, in all of its many forms, attempts to capture the success of the genetic recombination that occurs in sexual reproduction: the genes of multiple parents are mixed and matched to yield new combinations that are not present even in the parents.

A common form of combinatorial mutagenesis, though by no means the only form, is known as DNA shuffling or assembly PCR. In this method, as few as one or as many as 20 or more parental homologs of the gene of interest are obtained in isolated form, fragmented randomly with DNase I, and subjected to replication by PCR in the absence of primers. Prepared this way, the fragments prime each other at their overlapping regions, and successive rounds of PCR eventually generate full-length products. Ideally, therefore, the resulting chimeric progeny incorporate sequence elements that have already been selected for functionality in the parents, while suffering much fewer nonsense, frameshift, and other nonproductive mutations that are common with random mutagenesis methods. After expression and screening of the chimeric progeny, further rounds of combinatorial mutagenesis are often employed until no further improvements are obtained–this process is termed directed evolution.

Directed evolution has achieved extraordinary success in several industrial enzymes, resulting in 100- to over 10,000-fold enhancements of activity, altered substrate specificities, and stabilities to environmental conditions such as acidity, temperature, and solvent composition [24, 25]. Assembly PCR [26] as well as several other related combinatorial methods known by their acronyms as StEP, SHIPREC, ITCHY, SCRATCHY, CLERY, and RACHITT, are described in detail in the book, Directed Evolution Library Creation: Methods and Protocols [27].

2.5.4. Rational design. The mystery surrounding the way in which a one- dimensional sequence of amino acids folds into an active three-dimensional (3D) enzyme has fascinated researchers for decades. Now, the understanding that has emerged from their work, summarized in several excellent books [28–30], in combination with the crystallization and structural analysis of a number of industrially important enzymes, is making the rational mutagenesis of such enzymes possible.

Rational design typically begins with the computer-based structural representation of a protein of interest, ideally informed by high-resolution, 3D structures of the protein or near relatives obtained by x-ray crystallography or nuclear magnetic resonance (NMR) spectroscopy. Structures of enzyme-substrate, enzyme-product, and enzyme-inhibitor complexes are especially useful, because they identify the amino acids that comprise the catalytic part of the enzyme, or the active site. However, simulation is also possible with knowledge of only the primary (sequence) structure, albeit at lower fidelity, based on similarity to homologous proteins and current understanding of protein folding kinetics and thermodynamics. Extensive protein sequence and structure databases exist and are freely available, most notably including Swiss- Prot and its supplement, TrEMBL, found online at http://us.expasy.org/sprot/. Secondary and tertiary structure predictions may then be generated by the following homology modeling software: SWISS-MODEL, an automated protein modeling server at the GlaxoWellcome Experimental Research Station in Geneva, Switzerland (free online at http://swissmodel.expasy.org/); PredictProtein, an online service for sequence analysis and structure prediction maintained by the European Molecular Biology Laboratory - Heidelberg (EMBL) and the Columbia University Bioinformatics Center (CUBIC) at http://www.emblheidelberg.de/predictprotein/predictprotein.html; and Modeller, maintained by the University of California at San Francisco and also free of charge to academic researchers at http://salilab.org/modeller/modeller.html.

Once the protein of interest has a structural representation, the most promising sites for modification can be predicted, often computationally. These are typically close to the active site and accommodate or stabilize the proposed reaction mechanism. They are also in the binding pocket for the substrate, or make important structural contributions to the enzyme. By site- directed mutagenesis of the DNA that encodes these amino acids, in which a common approach is to allow oligonucleotides bearing the desired mutation(s) to prime PCR amplification reactions at the locus of interest, other residues can be substituted in their place(s) [31, 21]. Amino acids are usually exchanged with one of the other 19 natural residues, but it has recently become fashionable to incorporate so-called designer amino acids into proteins as well. The mutated gene is then introduced into a suitable organism and expressed, and the new protein product is partially or completely purified to reveal whether the amino acid substitution(s) have had the desired effect. This process is typically iterative, with multiple rounds of mutation and evaluation [32].

Alternatively, a 3D protein structure can be compared with those of related proteins that vary in the parameter of interest, either by models listed above or by a number of other homology modeling programs and online servers (see listing of those currently available at http://ncisgi.ncifcrf.gov/~ravichas/HomMod/). Comparisons among related proteins with variation in thermostability, for example, have revealed that higher stability generally correlates with greater proline, arginine, and tyrosine content; lower asparagine, glutamate, cysteine, and serine content; increased numbers of salt bridges and stabilizing hydrogen bonds; and a larger fraction of residues in alpha helices [33]. Unfortunately, however, the effect of a particular amino acid substitution on a protein's thermostability has not proven to be highly predictable with current understanding [34], typical of the situation with other parameters of interest as well. This limits speed and throughput, which then limits the sequence space amenable to testing. Nevertheless, the understanding of factors influencing protein folding is increasing rapidly, supported in great part by funding for basic research, and is expected to continually enhance efforts directed toward enzyme rational design [32].

Rational approaches have also entered an era of de novo design in both enzyme active sites and catalytic antibodies. In one method, the high-energy state of a reaction is modeled with a protein or antibody side chain geometrically oriented for catalysis. Then, a library of rotamers, or low-energy side chain conformations, is generated and the novel active sites are tested for optimal fit with a carrier or scaffold protein. This approach has been demonstrated successfully in the design of a novel active site for ester hydrolysis within the otherwise inert protein, thioredoxin [35]. This promises interesting future results for activities of industrial interest [36].

Rational design is clearly a time- and information-intensive process, and our understanding of enzymes is far behind that of small molecules. Still, it has succeeded in the reconfiguration of substrate specificities in oxidoreductases, hydrolases, transferases, and DNases; the alteration of cofactor requirements; the inversion of reaction stereochemistry; and the enhancement of enzyme stabilities under various industrial conditions, as well as the introduction of novel catalytic activities into existing templates [32]. In addition, sufficient structural information is accumulating for industrially important enzymes, including lipases and cellulases as well as hydrogenases, that rational design must be considered as one of the promising molecular techniques available for bioenergy development.

2.5.5. Analytical mutagenesis. Mutagenesis is also frequently employed to lend insight into the functions of metabolic pathways, specific genes, or specific regions within a gene, rather than to create genes with improved function. For these purposes, mutagenic methods are employed that frequently eliminate the function of a gene altogether. Transposon insertional mutagenesis is quite popular for this purpose, acting by the random insertion of large segments of DNA (the transposons) into a genome. Mutants lacking a function of interest are then assumed to have received an insertion within a gene essential to that function, and transposon sequences can then be used to locate the insertion. Sequences flanking the insertion are then investigated, ideally with the help of annotated genomic databases, to reveal the likely function of the interrupted gene [21].

If the identity of a gene is known, but the regions of amino acids that are most essential to the activity of the encoded protein are of interest, a technique known as linker-scanning mutagenesis may be used to probe the structure-function relationships among the amino acids.

In one approach to this technique, 10 to 15 base pair oligonucleotides are inserted into the gene sequence at random; alternatively, the mutagenesis may be designed such that internal deletions occur as well. Insertions or deletions that disrupt the function of the protein are assumed to occur in regions essential to the correct folding of the protein, while those that are tolerated are assumed to occur on the surface or in less-essential regions. In this way, the functional regions of an enzyme can be mapped with high resolution [1, 21, 31, 37].

Variations on the above-mentioned analytical mutagenesis methods are numerous, highly specialized, and evolving rapidly; for current and detailed information, the reader is referred to Current Protocols in Molecular Biology, updated quarterly [1].

2.6. Proteomics

The cellular proteome is the complete set of proteins found in a particular cell type under a specific set of conditions. The complete proteome, in turn, is the complete set of proteins that may be synthesized by the set of cellular proteomes, or the approximate protein analog of the genome. A cell's proteome typically possesses a much larger number of elements than does its genome, due to alternative processing of gene products and post-translational modifications such as glycosylation and phosphorylation. It is also more complex than the genome, in that it incorporates functional interactions among distinct proteins, and it is dynamic, in comparison to the generally static genome [38].

Proteomics is the study of proteomes, particularly including proteome structure and function, and includes protein separation, identification, quantification, and analysis of protein sequences, structures, modifications, and interactions. It therefore encompasses the development of technologies used in protein separation and structural analysis, such as two-dimensional (2D) electrophoresis and mass spectrometry, as well as the study of protein-protein and protein-DNA interactions that influence the synthesis of other proteins [38].

Proteomics is a recent addition to the landscape of biological research and represents an expansion in biological thought from concentration upon individual proteins and protein assemblages to large systems of interacting proteins. Proteomics is directly related to metabolic engineering, discussed below, in that it considers systems of enzymes interacting in metabolic pathways, as well as those interacting to govern cell division, cell signaling and response to environmental conditions, and transcriptional regulation that in turn determines the composition of the proteome itself [39, 40].

Like genomics, proteomics has emerged from the rapidly expanding set of databases, for example, Swiss-Prot, http://us.expasy.org/sprot/; Protein Data Bank, http://www.rcsb.org/pdb/) and data-mining tools as an application of bioinformatics that demands powerful computational resources and promises insights on greater scales of cellular complexity than have previously been possible [40, 41].

Since biotechnology relies directly upon the ability to control a cell's proteome, particularly including optimization of the balance between metabolic pathways that synthesize desired products and other pathways that maintain cell vigor, advances in proteomics will directly benefit all bioengineering endeavors.

2.7. Pathway Engineering

Metabolic pathway engineering integrates the approaches and technologies described above and has the fundamental goals of modifying biosynthetic pathways, assessing the physiological outcomes of the genetic modifications, and using the resulting information to improve further the pathways in question [42]. The optimization of entire metabolic networks, rather than individual enzymes, is often necessary because kinetic control is frequently distributed throughout a pathway rather than concentrated in a single reaction. When levels of a particular enzyme are altered, the fluxes not only of its direct product(s) and substrates are altered, but also of metabolites in related pathways linked to the pathway of interest through regulatory and common-substrate relationships. As a result, multiple points of intervention, frequently requiring fairly small changes in enzyme activity, may be required to achieve the desired metabolic changes [42, 43].

Pathway engineering is primarily directed toward one of several distinct goals. The first of these is the elucidation of pathways of interest, involving both identification of component reactions as well as reaction and/or transport bottlenecks [42]. Numerous mathematical tools are being brought to bear on this goal, in combination with genomic and microarray-generated expression data, and have had great success in elucidating the structures of metabolic pathways and distribution of kinetic control within them [42]. Two recent breakthroughs in this area are found in the development of a new kinetics format, termed linear log kinetics, which has proven remarkably accurate in describing intracellular kinetic behavior of metabolic networks, and the second is the development of a conceptual and experimental framework known as FANCY for elucidation of gene function by analysis of the metabolome, or total metabolite composition of a cell [43]. Additional promising in silico metabolic pathway modeling approaches interpret and predict cellular functions within the extremes of allowable possibilities, followed by the use of biochemical rationales to select the most reasonable behaviors [43]. This quantitative analysis of pathways leads to the understanding necessary, in turn, to target specific promising genetic modifications [42].

Another primary goal of pathway engineering, enabled by the first, is the modification of a pathway such that preferred substrates can be used, where preferred substrates are either less expensive, more widely available, or more environmentally friendly than the conventional substrates. Increased attention is being given to the use of renewable resources for the synthesis of specialty and commodity chemicals by so-called green processes. In addition, biomass- derived substrates are among the most widely available renewable resources–those generated from agriculture and municipal, agricultural, and forest wastes, among others.

The majority of inexpensive biomass is composed of lignocellulose. However, this contains a significant proportion of less-readily fermented five-carbon sugars in combination with the readily utilized six-carbon sugars. The great promise of pathway engineering for facilitating utilization of renewable agricultural materials is revealed by the development of metabolic pathways for the use of biomass-derived mixed sugars for the production of ethanol, described further in Chapter IV [43].

A third goal of pathway engineering is the development of pathways for the synthesis of novel chemical structures, particularly antibiotics, carotenoids, and polyhydroxyalkanoates, the latter of which are described further in Chapter III.

These efforts involve futuristic approaches, such as combining genes from different pathways and/or different organisms; inserting genes into a pathway or deleting genes from a pathway; combining modules derived from different multidomain enzymes to form new enzymes with novel catalytic activities; and engineering enzymes (e.g., by directed evolution) with new substrate specificities or catalytic activities in a pathway such that new and even non-natural substrates can be introduced, yielding novel products [44].

Pathway engineering has been greatly facilitated by genomic endeavors, which have provided access to sequence data not only of structural genes, but also of the genetic control elements central to the transcriptional regulation of genes, pathways, and groups of related pathways [42]. Gene expression databases have been equally valuable. In particular, the DNA microarray-based analysis of expression patterns characteristic of different physiological states, together with the characterization of transcriptional and post-translational controls within metabolic networks, have helped to identify key genetic targets for improvement of the desired biocatalysis. Microarray technology is also contributing to the elucidation of pathways by validating gene function, determining whether proteins are membrane-bound or cytosolic, and characterizing DNA-binding proteins [42].

Pathway elucidation is also highly dependent upon the ability to measure accurately the fluxes of metabolites within it. Considering the large numbers of metabolites present in a cell at any time, and the typically low concentrations of pathway intermediates, such measurements represent considerable challenges. To obtain intracellular flux data, Carbon 13 (^{13}C)-labeling of specific atoms within substrates and isotopomers (isotopic isomers) of intermediates are frequently used in combination with ^{13}C-NMR, 2D-NMR, or mass spectrometry. Alternatively, if the bioreaction sequence under investigation does not involve Carbon-Carbon (C—C) bond cleavage, uniformly Carbon 14(^{14}C)-labeled tracers can be used to reveal fluxes by tracking depletion of the radiolabeled tracer pulsed into the metabolite pool [42, 43].

While most pathway engineering successes are concentrated in pharmaceuticals and other medical applications, several examples demonstrate the great potential of pathway engineering in pollution prevention as well. Among these, one of the most famous is the development of a pathway to synthesize the blue dye indigo from glucose in bacteria, which avoids use of aromatic precursors and generation of aromatic wastes necessary in the conventional chemical synthesis. In the indigo pathway engineering process, the tryptophan synthesis pathway was first extended to indigo with the addition of one enzyme, naphthalene dioxygenase.

This addition resulted in the attenuation of activity of another important synthase by the conversion of the intermediate indoxyl to indigo, which required the increase of the gene dosage of that enzyme, the increase of its substrate availability, and the inactivation of a competing enzyme.

Finally, to diminish the content of indirubin, a product of a non-enzymatic side reaction in the conversion of indoxyl to indigo, in the final product, an isatin hydrolase was introduced. This successful example clearly illustrates the unexpected effects of altering a native pathway and the great effort needed to compensate for them to achieve the desired catalysis [45, 43].

3. Research Priorities

Current research in genetic engineering platform technologies is proceeding at an almost incomprehensibly rapid pace under the impetus of medical and basic biological research goals. While environmental biotechnology has much to gain from advances in these research goals, the support in these technologies and pace of progress are already sufficiently great, but funding for environmental goals is limited such that agencies with primarily environmental goals are encouraged to direct their support toward research priorities in other areas.

REFERENCES

[1] Ausubel, F. M., R. Brent, R. E. Kingston, D. D. Moore, J. G. Seidman, J. A. Smith, and K. Struhl, Eds. (1988 [updated quarterly]). *Current Protocols in Molecular Biology*, Ringbound Edition. Greene Publishing Associates.

[2] American Society of Human Genetics (2005). *The Meaning of Genetics as a Science*, http://genetics.faseb.org/genetics/ashg/educ/001.shtml.

[3] Benson, D. A., I. Karsch-Mizrachi, D. J. Lipman, J. Ostell, and D. L. Wheeler (2004). *GenBank: update*, Nucleic Acids Res 32:D23-D26.

[4] Chakrabarty, A. M. (2003). *Environmental biotechnology in the postgenomics era*, Biotech Adv 22:3-8.

[5] Frazier, M. E., G. M. Johnson, D. G. Thomassen, C. E. Oliver, and A. Patrinos (2003). *Realizing the potential of the genome revolution: the Genomes to Life Program*, Science 300:290-293.

[6] Glick, B. R., and J. J. Pasternack (2003). *Molecular Biotechnology*. ASM Press.

[7] De Boer, H. A., L. J. Comstock, and M. Vasser (1983). *The tac promoter: a functional hybrid derived from the trp and lac promoters*, Proc Natl Acad Sci 80:21-25.

[8] Fernandez, J., and J. Hoeffler, Eds. (1999). *Gene Expression Systems: using Nature for the Art of Expression*. Academic Press.

[9] Baneyx, F., Ed. (2004). *Protein Expression Technologies: Current Status and Future Trends*. Horizon Bioscience.

[10] Manoharan, M. (2004). *RNA interference and chemically modified small interfering RNAs*, Curr Opin Chem Biol 8:570-579.

[11] Mocellin, S., and M. Provenzano (2004). *RNA interference: learning gene knockdown from cell physiology*, J Transl Med 2:39.

[12] Cummings, C. A., and D. A. Relman (2000). *Using DNA microarrays to study host-microbe interactions*, Emerg Infect Dis 6:513-525.

[13] Synthegen (2005). *Table of Fluorescent Dyes, Synthegen Corporation*, http://www.synthegen.com/Action.lasso?-Response=/products/fluorescent/table.lasso&-Token. SortColumn=sort _order&-Nothing.

[14] Gershon, D. (2002). *Microarray technology: an array of opportunities*, Nature 416:885-891.

[15] Goldsmith, Z. G., and N. Dhanasekaran (2004). *The microrevolution: applications and impacts of microarray technology on molecular biology and medicine (review)*, Int J Mol Med 13:483-495.

[16] Friedberg, E. C., and G. C. Walker (1995). *DNA Repair and Mutagenesis*. ASM Press, Washington, D.C.

[17] Kodym, A., and R. Afza (2003). *Physical and chemical mutagenesis*, Methods Mol Biol 236:189-204.

[18] Hartwell, L. H., L. Hood, M. L. Goldberg, A. E. Reynolds, L. M. Silver, and R. C. Veres (2004). *Genetics: From Genes to Genomes,* 2nd Edition. McGraw Hill.

[19] Østergaard, L., and M. F. Yanofsky (2004). *Techniques for molecular analysis: establishing gene function by mutagenesis in Arabidopsis thaliana*, Plant J 39:682.

[20] Tobin, M. B., C. Gustafsson, and G. W. Huisman (2000). *Directed evolution: the "rational" basis for "irrational" design*, Curr Opin Struc Biol 10:42 1-427.

[21] Braman, J., Ed. (2002). *In Vitro Mutagenesis Protocols*, 2nd Edition. Humana Press, Totowa, NJ.

[22] Cirino, P. C., K. M. Mayer, and D. Umeno (2003). *Generating mutant libraries using error-prone PCR*, Methods Mol Biol 231:3-9.

[23] Biles, B. D., and B. A. Connolly (2004). *Low-fidelity* Pyrococcus furiosus *DNA polymerase mutants useful in error-prone PCR*, Nucleic Acids Res 32:e 176.

[24] Crameri, A., S.-A. Raillard, E. Bermudez, and W. P. C. Stemmer (1998). *DNA shuffling of a family of genes from diverse species accelerates directed evolution*, Nature 391:288-291.

[25] Christians, F. C., L. Scapozza, A. Crameri, G. Folkers, and W. P. C. Stemmer (1999). *Directed evolution of thymidine kinase for AZT phosphorylation using DNA family shuffling*, Nat Biotechnol 17:259-264.

[26] Joern, J. (2003). *DNA Shuffling, in Directed Evolution Library Creation: Methods and Protocols* (F. H. Arnold and G. Georgiou, Eds). Humana Press, pp 85-90.

[27] Arnold, F. H., and G. Georgiou, Eds. (2003). *Directed Evolution Library Creation: Methods and Protocols*. Humana Press, Totowa, NJ.

[28] Pain, R. H. (2000). *Mechanisms of Protein Folding*, 2nd Edition. Oxford University Press.

[29] Schlick, T. (2002). *Molecular Modeling and Simulation.* Springer.

[30] Buchner, J., and T. Kiefhaber, Eds. (2005). Protein Folding Handbook. John Wiley & Sons.

[31] Trower, M. K., Ed. (1996). In Vitro *Mutagenesis Protocols*, 1st Edition. Humana Press, Totowa, NJ.

[32] Weissman, K. (2004). *Rational or Random?*, Chemistry World 1(7): July, http://www.rsc.org/chemistryworld/.

[33] Lehmann, M., and M. Wyss (2001). *Engineering proteins for thermostability: the use of sequence alignments versus rational design and directed evolution*, Curr Opin Biotechnol 12:371-375.

[34] Spector, S., A. Wang, S. A. Carp, J. Robblee, Z. S. Hendsch, R. Fairman, B. Tidor, and D. P. Raleigh (2000). *Rational modification of protein stability by the mutation of charged surface residues,* Biochem 39:872-879.

[35] Bolon, D. N., and S. L. Mayo (2001). *Enzyme-like proteins by computational protein design*, Proc Natl Acad Sci 98:14274-14279.

[36] Bolon, D. N., C. A. Voigt, and S. L. Mayo (2002). De novo *design of biocatalysts,* Curr Opin Chem Biol 6:125-129.

[37] McKnight, S. L., and R. Kingsbury (1982). *Transcriptional control signals of a eukaryotic protein-coding gene*, Science 217:316-324.

[38] Wikipedia (2005). *Proteomics*, http://en.wikipedia.org/wiki/Proteomics.

[39] Liebler, D. C. (2001). *Introduction to Proteomics: Tools for the New Biology*. Humana Press, Totowa, NJ.

[40] Twyman, R. M. (2004). *Principles of Proteomics*. BIOS Scientific Publishers, New York.

[41] Westermeier, R., and T. Naven (2002). *Proteomics in Practice: A Laboratory Manual of Proteome Analysis*. Wiley-VCH, Weinheim.

[42] Stafford, D. E., and G. Stephanopoulos (2001). *Metabolic engineering as an integrating platform for strain development*, Curr Opin Microbiol 4:336-340.

[43] Sanford, K., P. Soucaille, G. Whited, and G. Chotani (2002). *Genomics to fluxomics and physiomics—pathway engineering,* Curr Opin Microbiol 5:318-322.

[44] De Boer, A. L., and C. Schmidt-Dannert (2003). *Recent efforts in engineering microbial cells to produce new chemical compounds,* Curr Opin Chem Biol 7:273-278.

[45] Berry A., T. C. Dodge, M. Pepsin, and W. Weyler (2002). *Application of metabolic engineering to improve both the production and use of biotech indigo,* J Ind Microbiol Biotechnol 28:127-133.

B. BIOREACTOR ECHNOLOGIES [45]

1. Introduction

All bioreactors have in common the central purpose of maintaining ideal conditions for one or more species of microbes, such that the maximum desired activity is promoted. Bioreactors are essential elements in all industrial processes that make use of microbial growth and metabolism with the expectation that the future success of commercial biotechnology depends significantly on the advancement of bioreactor technology [1]. Basic bioreactor design has changed little, however, in the past 40 years, possibly due to an absence of sufficient market forces. Currently, more than 50 percent of all commercial bioreactors are used in the synthesis of low-volume, high-value products, such as pharmaceuticals, that may be synthesized profitably without stringent process optimization. In contrast, the productions of high-volume, low-value products such as biomaterials and fuels carry sufficiently low profit margins that a high degree of optimization is essential. Because the latter class of products has the greatest potential for positive environmental impact, bioreactor design and optimization represent important priorities in the advancement of bioengineering for pollution prevention.

1.1. Reactor Conditions

Successful design and operation of bioreactors requires optimization of numerous quantities, including such conditions as pH, substrate and product concentrations, cell density, oxygen concentration, and temperature. Understanding the roles of each of these in a particular bioprocess, as well as predicting and controlling their spatial and temporal variation within narrow limits throughout the bioreactor volume, is of central importance to reactor design. Consequently, primary goals of current engineering efforts include accurate modeling of bioprocesses, real-time sensing of internal reactor conditions, and the design of bioreactors in which conditions can be maintained as nearly uniform as possible.

1.2. Reactor Types

Numerous varieties of commercial-scale bioreactors are in use, with the majority categorized as unstirred vessels, stirred vessels, bubble columns, airlift reactors, membrane reactors, fluidized beds, or packed beds [2]. While each category presents a unique set of advantages and limitations, the considerations relevant to bioprocesses are quite similar throughout.

Large-scale industrial fermentations are typically conducted as fed-batch processes, involving intermittent supply of substrate(s), which utilizes substrates that are as inexpensive as possible to yield a product density as high as possible. Fed-batch processes offer several advantages over traditional batch operations: first, they avoid catabolite repression effects, which occurs when excess quantities of a substrate such as glucose can redirect metabolic energy and carbon flow; second, they avoid overflow metabolism, which occurs when substrate in excess of the capacity of a particular pathway is converted to undesirable co-metabolites; and finally, they minimize product-inhibition effects, which occurs when accumulation of a product can inhibit further progress of a desired reaction [3].

An example of this is the application of the fed-batch process to the culture of baker's yeast, Saccharomyces cerevisiae. The yeast effectively diminishes glucose repression and overflow metabolism occurs leading to ethanol formation [4]. Similarly, in the industrially important Escherichia coli, fed-batch processing diminishes the production of acetate through overflow metabolism and thereby allows greater energy flow toward the production of desired recombinant proteins [5].

1.3. Reactor Models

The design, optimization, and real-time control of bioreactors depend on the availability of good process models. The central role of these models is the establishment of quantitative relationships between kinetic processes (those involving enzymes) and transport processes (those involving the flow and distribution of substrates, cells, products, and wastes within the bioreactor).

The former involve enzyme rates and specificity (or selectivity) characteristics, as well as kinetic characteristics of microbial growth, substrate uptake, and longevity, which are revealed through basic biochemical research. The latter, in turn, involve considerations of fluid mechanics and heat and mass transfer that require considerable computational resources.

Accurate models become especially important when bioprocesses must be scaled up from bench to larger commercial volumes without losses of product quality, yield, or process stability. Heat removal and gas-liquid mass transfer limitations, for example, can present great scale-up challenges that are addressed much more readily with the assistance of accurate models.

2. State of the Science

2.1. Kinetic Models

Kinetic expressions describe quantitative relationships among enzyme concentrations, substrate concentrations, product concentrations, and rates of product generation. Reliable kinetic expressions are highly advantageous to the design of cost-effective bioprocesses, especially in the production of comparatively low-value, high-volume products such as biofuels and biomaterials. In these cases, in which small losses of substrates to undesired metabolic byproducts can render a process economically impractical, kinetic models are essential to the achievement of high substrate conversion efficiency.

2.1.1. Metabolic flux analysis. In many biosynthetic pathways of interest, simple understanding of metabolite transformation kinetics (substrate consumption rates, intermediate lifetimes, and product production rates) is sufficient for reactor design, causing understanding of individual enzyme kinetics to be unnecessary. In these cases, an approach known as metabolic flux analysis, or MFA, is convenient to describe the flow of each metabolite through nodes, or enzymatic transformations, throughout an entire biochemical network. The result is a comprehensive flux map that shows and quantifies all major anabolic (synthetic) and catabolic (degradative) processes of interest, as well as points of sensitivity to perturbation of metabolite or enzyme concentrations, or flux control coefficients, within an organism [6, 7]. Once such a mathematical framework has been established experimentally, fluxes of particular metabolites can be predicted over ranges of conditions and flux maps can be compared among different organisms and growth conditions to reveal optimal process conditions. In addition, flux maps and comparisons among different flux maps may indicate possible targets for genetic modifications within the cultivated organism, reveal the outcome of genetic manipulations that have already been performed, or yield more basic insight regarding cellular energy metabolism [8].

In the initial development of a metabolic flux map, all theoretically possible enzyme reactions in the network are considered; postulation of a map appropriate to a particular process then involves the designation of some reactions as dormant. Metabolic flux analysis experiments are typically performed in stirred-tank reactors (STRs), also known as chemostats, to establish steady-state conditions. Concentrations of substrates, intermediates (where possible), and products are then monitored over time to reveal their interdependences, and the flux map is refined based on the results. Highly reproducible data are prerequisite for the confirmation of a postulated metabolic network, however, and until pathways have been rigorously established, the uniqueness of the proposed network is typically viewed with caution [9].

For the analysis of large metabolic networks, the use of isotopically labeled substrates (e.g., $^{13}CH3COO-$) is often required to trace the flows of compounds present at low concentrations. The great amount of data generated by such experiments can lead to sizable computational challenges, because fluxes through individual pathways must then be calculated by solving large sets of non-linear equations. Fortunately, new quasi-linear algebra methods have been developed to calculate fluxes from large data sets and, importantly, to estimate sensitivities to measurement errors [10]. A number of systematic descriptions of metabolic pathways for E. coli have been developed using these methods [11-13].

2.1.2. Enzymatic kinetic models. Detailed kinetic models, involving descriptions of enzyme characteristics, fall into three broad categories. The most detailed and comprehensive are referred to as mechanistic or structured models. In such models, rate expressions are proposed that attempt to describe the mechanistic role of each enzyme in the metabolic pathway. A fundamental structured model also includes information about characteristic microbial cell dimensions and effects of rate-limiting mass transfer across cell walls.

In a structured model, the limiting substrate participates in the first reaction pathway; the product of this reaction participates as a reactant in the next pathway, and so on to capture the full cascade of transformations as well as regulatory feedback reactions. Certain structured models have been successful in predicting important effects, such as the differential uptake of two separate substrates. This confirms the value of such models when they can be properly validated [14]. Currently, however, the use of structured models is limited by the lack of rigorous verification, which in turn results from the considerable experimental challenges involved in determining numerous enzymatic parameter values. Accordingly, utilization of such models for design purposes is restricted to the range of parameters that can be validated by experimental data [15].

The second, probably most widely-used set of kinetics models comprises those that assume the presence of a single growth-limiting substrate. Among these so-called black-box models, the most prominent is the Monod equation, in which the biomass specific growth rate, μ, is related to the substrate concentration, S, in a nonlinear form (Equation 1). The parameters KS and imax, describing the half-saturation constant and maximum specific growth rate, respectively, are determined experimentally [16].

$$\mu = \mu_{max} \frac{[S]}{K_S + [S]}$$

(Equation 1)

Extensions of the Monod equation to capture effects of metabolite repression and other inhibitory and/or limiting effects are available. Black-box models can be extremely useful in their estimations of overall fermentation behavior in early stages of process design. At the same time, they lack the level of detail often necessary for precise optimization of bioreactor function [15].

Another class of black-box models comprises those based on artificial neural networks (ANNs). ANNs structure is characterized by the total number of nodes, responsiveness of each node to an average input, and the response function for each node [17].

Model parameters for ANNs are termed weights, and the process of determining weights from experimental data is called the training procedure. While utility of these models is again restricted to the region over which experimental data are available, they offer the valuable ability to accommodate increased metabolic complexity compared to limiting-substrate models [18].

Finally, so-called gray-box modeling is also applied to bioreactor design, referring to strategies that attempt to combine both fundamental knowledge and empirical data to obtain models of moderate complexity with qualitative behavior in reasonable agreement with experimental data. This class of models is best represented, both conceptually and industrially, in the form of fuzzy-rule systems. Fuzzy-rule theory was developed by Zadeh [19, 20] and has become increasingly important in practical process modeling and control. Relationships between fuzzy variables are, in turn, formulated using fuzzy logic operators to reflect the common practices used by operators in everyday bioreactor operation. Numerous kinetic expressions for bioreactors have been formulated using this approach [21, 22], and fuzzy modeling and control are now regarded as promising methods for automating industrial bioprocesses in which experienced operators play significant roles in their successful operation [22].

2.2. Transport Phenomena and Models

Transport equations predict gradients of dissolved substances, temperature, etc. within fluids. They are based on principles of conservation of mass, momentum, and energy. When applied to bioreactor operation, they can be used both to discover and explain phenomena of interest, as well as to guide bioreactor design and scale-up. Because of the central importance of reactor uniformity, transport modeling is frequently used to address issues of mixing, mass transfer between gases and liquids, heat removal and/or maintenance of optimal microbial growth temperatures, and biofilm formation on reactor surfaces. The distribution of fluid velocities, or velocity profile, within a bioreactor is especially important in the calculation of gas transport and heat transfer patterns as well as shear stresses that influence locations of biofilm formation.

The complete simulation of mass and energy transport throughout all parts of a reactor, showing consequences of adjusting design variables, would be ideal for reactor design. While the typical two-phase gas-liquid media composition, locally turbulent flows, and limitations of kinetic models greatly complicate the calculations involved in traditional transport models, new approaches are being developed that hold great promise. In recent years, computational fluid dynamics (CFD) in particular has enabled the capture of salient features in bioreactors. For example, the simulation of a bubble column fermenter is shown in Figure 2. Instantaneous values are shown on the left, while time averages are shown on the right; velocity profiles are plotted with arrows while oxygen distributions are shown in color. For these calculations, the two-phase gas-liquid system has been treated as a homogeneous medium with a variable density that depends on the gas retention in the column, resulting in visibly turbulent flow at short timescales. In this example, quantification of the spatial variation of the oxygen combined with a kinetic model based on oxygen provided the rate of product generation.

Other reactor types present even more challenging modeling problems. For example, in STRs, involving impellers that rotate rapidly through a two-phase flow, boundary conditions on the transport equations require the use of moving boundary computational grids [23, 24]. Despite the challenges, describing and predicting details of mass and energy flow throughout bioreactors mathematically–including the realistic representation of fluids that exhibit non-Newtonian behavior due to the suspension of cells and particulates–remain important goals due to their great potential to facilitate reactor design [25].

2.3. On-line Sensing

Real-time sensing of bioreactor conditions, involving spatially resolved measurements of fluid velocities and reaction components, is essential for both the experimental validation of bioreactor models and for monitoring of ongoing performance. Even the most perfectly-designed and thoroughly-modeled bioreactor is expected to experience unforeseen conditions occasionally, particularly given the presence of mutable microorganisms, with the result that real-time or on-line sensing of bioreactor conditions during operation is essential. On-line sensing is also important, of course, in validating models during development. Especially important is the potential of on-line sensing to allow precise, automated, feedback control devices to maintain reactor homeostasis.

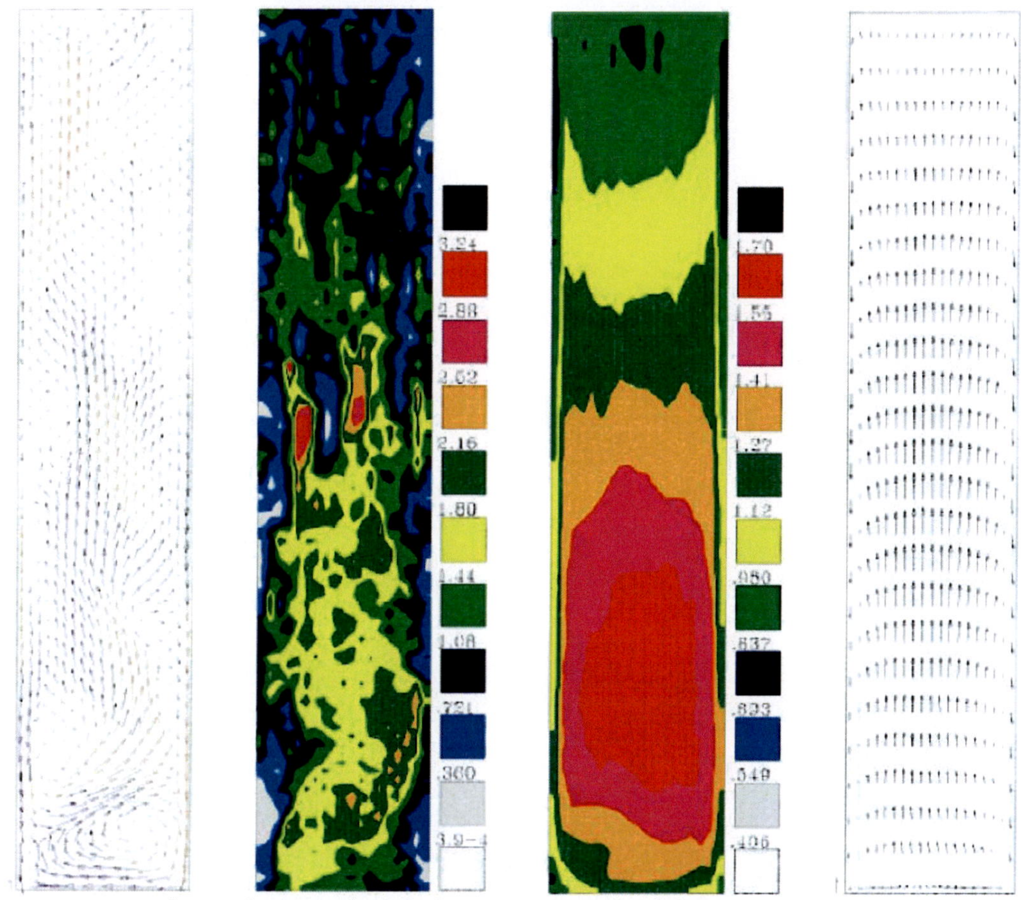

Instantaneous results are shown on the left, time averages on the right.

Figure 2. Distribution of fluid velocity (arrows) and oxygen distribution (color).

Conventional reactor sensors are often not ideal for bioprocess measurements, however, due to their vulnerabilities to interference by biofilm growth, inabilities to resolve overlapping signals generated in complex culture media, and inabilities to be sterilized to avoid contaminating culture media, for example [26, 27], and ongoing research into sensor design is important to the future of bioprocessing.

At the same time, sufficient progress has been made in several key areas that a number of important variables may be monitored adequately. Dissolved oxygen and pH, for example, are frequently monitored on-line using electrochemical sensors contained within steam-sterilizable glass electrodes [28]. Dissolved oxygen may also be measured by a recently-commercialized method based on fluorescence quenching [29]; available sensors are described at www.oceanoptics.com and www.fluorometrix.com. On-line measurements of cell mass are also desirable and can now be made indirectly by electrical or optical means: capacitance and permittivity measurements, for example, provide electrical quantitation [30], while light absorbance, light scattering, or a combination of the two provide optical indications of cell mass [31].

Carbon dioxide, a waste product of cell respiration, is an important indicator of bioreactor status and can be measured on-line by means of sterilizable electrode sensors, optical sensors, and sensors based on gas permeation through selective polymer membranes [32, 33], serving as the basis for process control loops [33].

Several spectroscopic techniques are also available to quantify the presence of numerous organic and inorganic species simultaneously, due to recent advances in optics and computing. An example is shown in Figure 3, in which a noninvasive sensor has measured the whole-cell biotransformation of L-serine and indole to tryptophan [34]. This approach is also applicable to the processing of sugar beet molasses at an industrial scale [35]. Glucose, fructose, glutamine, ammonia, CO_2, and phosphate are among the many compounds that can be measured by either near or mid-infrared spectroscopy. Recent developments of improved, low-cost optical sensors are also promising [27, 36, 37]. Further development for miniaturization, improved robustness, and sensitivity are expected [38, 39].

A separate approach is based on the attachment of the gene for green fluorescent protein onto a protein of interest present during manufacture [40]. This is particularly attractive if the tagged molecule is the product of interest, such that product concentration can be measured directly. A number of other color probes are also currently available, opening a promising avenue for real-time monitoring of the expression of multiple genes simultaneously [41].

Improved real-time sensing of bioreactor conditions is essential to model validation and process control during operations, and future bioreactors are expected to be massively instrumented to provide detailed real-time information of vital interest.

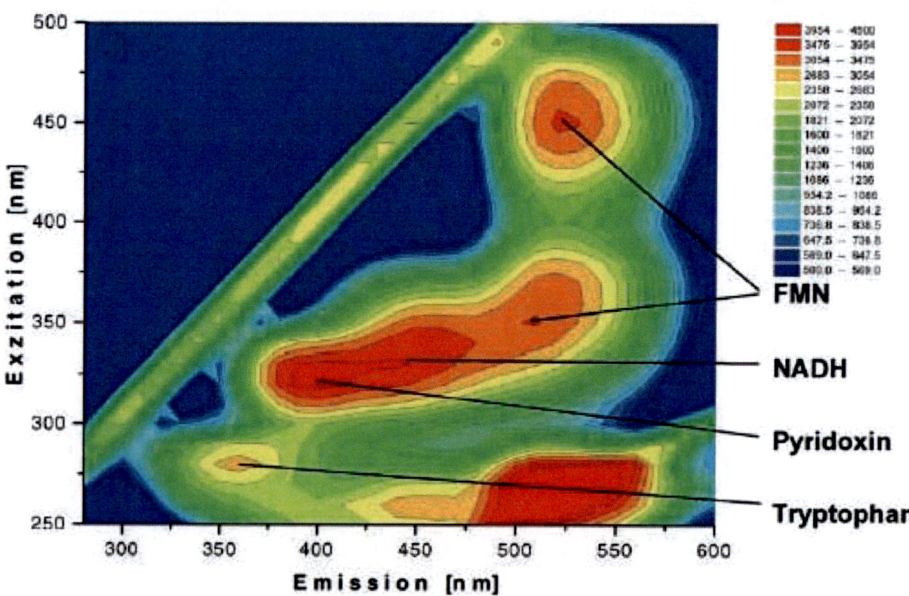

Figure 3. Two-dimensional fluorescence spectra of a fermentation broth [34].

Advances in sensing technologies are urgently needed, especially in the design of bioprocesses for commodity products where efficiency is paramount, and this area should be considered a top priority within bioengineering for pollution prevention.

3. Research Priorities

Because bioengineering for pollution prevention involves relatively low-value products, requiring optimal bioprocessing for commercial feasibility, improvements in bioreactor technology should be a high priority in general in this field. In addition, several areas are worthy of specific mention:

3.1. Sensing

Real-time sensing of gases and aqueous metabolites is of central importance because it allows or has the potential to facilitate model validation, development of descriptive kinetic expressions, and real-time process control based on sensor feedback alone and in combination with model predictions. Biosensors suitable for monitoring bioconversions in bioreactors have been previously identified as a bottleneck in the development of high-volume, low-cost processes [42], indicating that the development of promising emerging technologies should be encouraged in every instance possible.

3.2. Modeling

In process design and optimization, the utility of a mathematical model lies in its ability to predict the operating characteristics in regions for which experimental data do not exist. Present kinetic models generally do not allow such procedures in great detail. Detailed kinetic models are also useful for capturing the dynamic responses of bioreactors to external stimuli, a set of important concerns in process control. Accordingly, a promising area of investigation is the development of more-detailed structural models that capture the salient features of complete metabolic pathways through the integration of biochemistry, molecular biology, and computational techniques. In particular, new approaches to structural kinetic modeling that include transcriptional and post-translational regulatory effects are needed.

In addition, new developments are needed in computational methods to capture effects of turbulence in bioreactors, effects of shearing and mechanical stresses on cellular growth and death, and the non-Newtonian nature of cellular media. These must involve multi-scale modeling to capture details at small spatial scales and must also be able to transfer relevant information to lower-resolution models that describe greater volume and time scales. CFD models are now able to establish hydrodynamic profiles among different zones of a reactor and could serve as the basis for such models, describing concentration and temperature gradients both instantaneously and over time. Such models could be extended to bridge length, volume, and time scales, linking detailed calculations at smaller scales or in critical areas with lower- resolution models that track averaged quantities. Given recent and continuing increases in inexpensive computing power, such models could contribute greatly to reactor optimization. In addition, they have the potential to guide the scale-up of industrial processes by revealing important mass and heat transfer limitations as reactor configurations are changed.

3.3 Applications of Genetic Engineering

Certain challenges inherent in bioreactor operation can be greatly alleviated by the skillful application of genetic engineering technologies. For example, substrate and product-based inhibitions are common phenomena which occur when enzyme activities diminish in the presence of locally high concentrations of certain metabolites. Increasing reactor-mixing can often alleviate these problems, but diminishing the inhibitory mechanisms genetically can offer a Because bioengineering for pollution prevention involves relatively low-value products, requiring optimal bioproces sing for commercial feasibility, improvements in bioreactor more convenient solution. An example of this approach is presented by Agger and colleagues, who disrupted the gene responsible for glucose repression so that the glucose conversion rate did not decrease with increasing glucose concentrations that were, in turn, needed to work at high biomass concentrations [43]. The ability of genetic engineering to solve bioreactor-based problems offers a set of great opportunities for improvements in reactor productivity.

4. Commercialization

Commercial bioreactor usage is growing rapidly, as shown in Figure 4, largely due to increases in pharmaceutical and food-based biotransformations, which account for over half and approximately one-quarter of the total biotransformations, respectively [44]. An important issue for the commercialization of bioreactors for pollution prevention goals is simply capitalization: entry into established markets for fuels and materials will require large economies of scale, meaning that production facilities must be constructed on a large scale to be profitable.

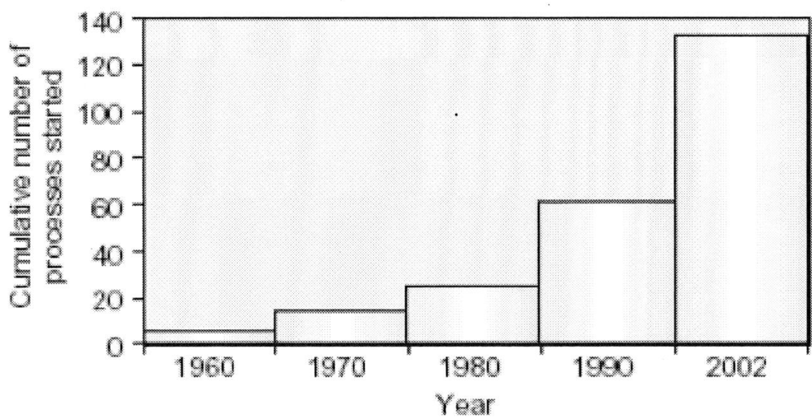

Figure 4. The growth of commercially practiced biotransformations [44]

At the same time, investors are typically reluctant to take risks with newer technologies, which results in bioreactors producing biofuels and biomaterials that may face significant initial barriers to market entry. The establishment of pilot plants for both biofuel and bioplastics production, described in subsequent sections, indicates that these barriers are nevertheless far from insurmountable.

REFERENCES

[1] Cooney, C. L. (1983). *Bioreactors: design and operation*, Science 19:728-740.

[2] Sittig, W. (1982). *The present state of fermentation reactors*, J Chem Tech Biotech 16: 16-21.

[3] Hochfeld, W. L. (1994). *Growth & Synthesis: Ferm enters, Bioreactors, & Biomolecular Synthesizers*. Interpharm Press.

[4] Rose, A. H. (1979). *History and scientific basis of large-scale production of biomass*, in Economic Microbiology (A. H. Rose, Ed.). Academic Press, London, pp 1-29.

[5] Majewski, R. A., and M. M. Domach (1990). *Simple constrained optimization view of acetate overflow in* E. coli, Biotechnol Bioeng 35:732-738.

[6] Liao, J. C., and E. N. Lightfoot (1988). *Characteristic reaction paths of biochemical reaction systems with time scale separation*, Biotechnol Bioeng 31:847-854.

[7] Delgado, J., and J. C. Liao (1992). *Determination of flux control coefficients from transient metabolite concentrations*, Biochem J 282:919-927.

[8] Stephanopoulos, G., A. Aristidou, and J. Nielsen (1998) *Metabolic Engineering: Principles and Methodologies*. Academic Press.

[9] Nielsen, J., and J. Villadsen (1994). *Bioreaction Engineering*. Plenum Press, New York.

[10] Weichert, W., M. Moellney, N. Iserman, M. Wurzel, and A. de Graaf (1999). *Biodirectional reaction steps in metabolic networks: III. Explicit solution and analysis of isotopomer labelling systems*, Biotechnol Bioeng 66:69-85.

[11] Varma, A., B. W. Boesch, and B. O. Palsson (1993). *Stochiometric interpretation of E. coli glucose catabolism under various oxygenation rates*, Appl Environ Microbiol 59:2465-2473.

[12] Varma, A., and B. O. Palsson (1993). *Metabolic capabilities of* E. coli *II. Optimal growth patterns*, J Theor Biol 165:503-522.

[13] Varma, A., and B. O. Palsson (1994). *Stochiometric flux balance models quantitatively predict growth and metabolic by-product secretion in wild-type* E. coli W 3110, Appl Environ Microbiol 60:3724-3731.

[14] Sonnleitner, B., and O. Kappeli (1986). *Growth of S. cerevisiae is controlled by its limited respiratory capacity: formulation and verification of a hypothesis*, Biotechnol Bioeng 28:927-937.

[15] Ishii, N., M. Robert, Y. Nakayama, A. Kanai, and M. Tomita (2004). *Toward large-scale modeling of the microbial cell for computer simulation*, J Biotechnol 113:281-294.

[16] Bitton, G. (1999). *Wastewater Microbiology*. Wiley-Liss.

[17] Bhat, N., and T. J. McAvoy (1990). *Use of neural nets for dynamic modelling and control of chemical process systems*, Comput Chem Eng 14:573-582.

[18] Gadkar, K. G., S. Mehra, and J. Gomes (2005). *On-line adaptation of neural networks for bioprocess control*, Comput Chem Eng 29:1047-1057.

[19] Zadeh, L. A. (1965). *Fuzzy Sets, Inform Contr* 8:338-353.

[20] Zadeh, L. A. (1973). *Outline of a new approach to the analysis of complex systems and decision processes*, IEEE Trans Sys Man Cybern 3:28-44.

[21] Lee, B., J. Yen, L. Yang, and J. C. Liao (1999). *Incorporating qualitative knowledge in enzyme kinetic models using fuzzy logic*, Biotechnol Bioeng 62:722-729.

[22] Horiuchia, J.-I. (2002). *Fuzzy modeling and control of biological processes: review,* J Biosci Bioeng 94:574-578.

[23] Reuss, M., and R. Bajpai (1991). *Stirred tank models - Measuring, modelling, and control, in Biotechnology* Vol 4. (H. J. Rehm, Ed.). VCH, Weinheim, pp 299-348.

[24] Reuss, M., S. Schmalzriedt, and M. Jenne (2000). *Application of Computational Fluid Dynamics (CFD) to modeling stirred tank bioreactors,* in Bioreaction Engineering (K. Schugerl and K. H. Bellgart, Eds.). Springer, Berlin, pp 207-246.

[25] Leib, T. M., C. J. Pereira, and J. Villadsen (2001). *Bioreactors: a chemical engineering perspective,* Chem Eng Sci 56:5485-5497.

[26] Gastrock, G., K. Lemke, and J. Metze (2001). *Sampling and monitoring in bioprocessing using microtechniques,* Rev Molec Biotechnol 82:123-135.

[27] Bellon-Maurel, V., O. Orliac, and P. Christen (2003). *Sensors and measurements in solid state fermentation: a review,* Process Biochem 38:881-896.

[28] Johnson, M. J., J. Borkowski, and C. Engblom (2000). *Steam sterilizable probes for dissolved oxygen measurement,* Biotechnol Bioeng 67:645-656.

[29] Bambot, S. B., R. Holavanahali, J. R. Lakowicz, G. Carter, and G. Rao (1994). *Phase fluorometric sterilizable optical oxygen sensor,* Biotechnol Bioeng 43:1139-1145.

[30] Zeiser, A., C. B. Elias, R. Voyer, B. Jardin, and A. A. Kamen (2000). *On-line monitoring of physiochemical parameters of insect cell cultures during the growth and infection process,* Biotechnol Prog 16:803-808.

[31] Janelt, G., N. Gerbsch, and R. Buchholz (2000). *A novel fiber optic probe for on-line monitoring of biomass concentrations,* Bioprocess Eng 22:275-279.

[32] Chang, Q., L. Randers-Eichhorn, J. R. Lakowicz, and G. Rao (1998). *Steam-sterilizable, florescence lifetime-based sensing film for dissolved carbon-dioxide,* Biotechnol Prog 14:326-33 1.

[33] Pattison, R. N., J. Swamy, B. Medenhall, C. Hwang, and B. T. Frohlich (2000). *Measurement and control of dissolved carbon dioxide in mammalian cell culture process using an in situ fiber optic chemical sensor,* Biotechnol Prog 16:769-774.

[34] Skibsted, E., C. Lindemann, C. Roca, and L. Olsson (2001). *On-line bioprocess monitoring with a multi-wavelength fluorescence sensor using multivariate calibration,* J Biotechnol 88:47-57.

[35] McGovern, A. C., D. Broadhurst, J. Taylor, N. Kaderbhai, M. K. Winson, D. A. Small, J. J. Rowland, D. B. Kell, and R. Goodacre (2002). *Monitoring of complex industrial bioprocesses for metabolite concentrations using modern spectroscopies and machine learning: application to gibberellic acid production,* Biotechnol Bioeng 78:527-53 8.

[36] Bashir, R. (2004). *BioMEMS: state-of-the-art in detection, opportunities and prospects,* Adv Drug Deliv Sys 56:1565-1586.

[37] Pons, M.-N., S. Le Bonte, and O. Potier (2004). *Spectral analysis and fingerprinting for biomedia characterisation,* J Biotechnol 113:211-230.

[38] Doak, D. L., and J. A. Phillips (1999). *In situ monitoring of an E. coli fermentation using a diamond composition ATR probe and mid-infrared spectroscopy,* Biotechnol Prog 15:529-539.

[39] Vaidyanathan, S., A. Arnold, L. Matheson, P. Mohan, G. Macaloney, B. McNeil, and L. M. Harvey (2000). *Critical evaluation of models developed for monitoring an*

industrialsubmerged bioprocess for antibiotic production using near-infrared spectroscopy, Biotechnol Prog 16:1098-1105.

[40] Chen, H. J., M. P. DeLisa, H. L. Cha, W. A. Weigand, G. Rao, and W. E. Bentley (2000). *Framework for on-line optimization of recombinant protein expression in high-celldensity E. coli cultures using GFP-fusion monitoring*, Biotechnol Bioeng 69:275-285.

[41] Li, J., S. Wang, W. J. Van Dusen, L. D. Schultz, H. A. George, W. K. Herber, H. J. Chae, W. E. Bentley, and G. Rao (2000*). Green fluorescent protein in real-time studies of the GAL1 promoter*, Biotechnol Bioeng 70:187-196.

[42] Lubbert, A., and S. B. Jorgenson (2001). *Bioreactor performance: a more scientific approach for practice*, J Biotechnol 85:187-212.

[43] Agger, T., A. B. Spohr, and J. Nielsen (2001). *Alpha-amylase production in high cell density submerged cultivations of Aspergillus oryzae*, Appl Microbiol Biotechnol 55:8 1- 84.

[44] Straathof, A. J. J., S. Panke, and A. Schmid (2002). *The production of fine chemicals by biotransformations*, Curr Opin Biotechnol 13:548-556.

C. BIOSEPARATIONS AND BIOPROCESSING

1. Introduction

The bioproducts considered in this report, primarily including materials either made by living organisms or derived from biomass, typically require extraction from either a whole organism, such as a plant; from aqueous bioreactor media (also referred to as fermentation broth or culture media; or from downstream processing solutions. A prevalent challenge presented by bioreactor-based processes, in particular, is the generation of products within relatively dilute aqueous solutions. Bioreactor media usually must remain dilute, however, to prevent inhibition of enzyme activity by accumulated products and to prevent cell mortality due to accumulated wastes. Nevertheless, the design of bioreactors to allow higher solute concentrations while maintaining cell health and activity is worthy of considerable effort.

Traditional separation techniques are well-developed, widely practiced, and comprehensively described in standard references [1]. The techniques are categorized according to their fundamental mechanisms as physical (adsorption, crystallization, extraction, etc.), mechanical (filtration, centrifugation, etc.), thermal (distillation), or chemical (chemisorption, chromatography). Because separations often dominate the economics of biochemical processing, development of energy-efficient separation methods is of especially great importance to the commercial success of the low-value, high-volume bioproducts most relevant to pollution prevention.

Downstream processing of bioreactor contents involves special challenges due to the presence of biomass and non-product biomolecules such as proteins and sugars, which promote fouling (coverage with biofilms or clogging with bioparticles), and also due to the necessarily dilute nature of fermentation broths, which causes the use of energy-intensive distillation to be prohibitively expensive [1].

A generalized block diagram of downstream processing of bioreactor media is shown in Figure 5 [2]. During primary recovery, preliminary separation of solid (biomass) and aqueous (product) phases, as well as product extraction, occur. The fermentation broth is also reduced significantly in volume during this stage. Products are further concentrated during intermediate recovery, which also involves, in some cases, the redissolution of materials before further concentration is possible. In the final purification stages, a series of polishing steps are used to raise product purity to the final specifications.

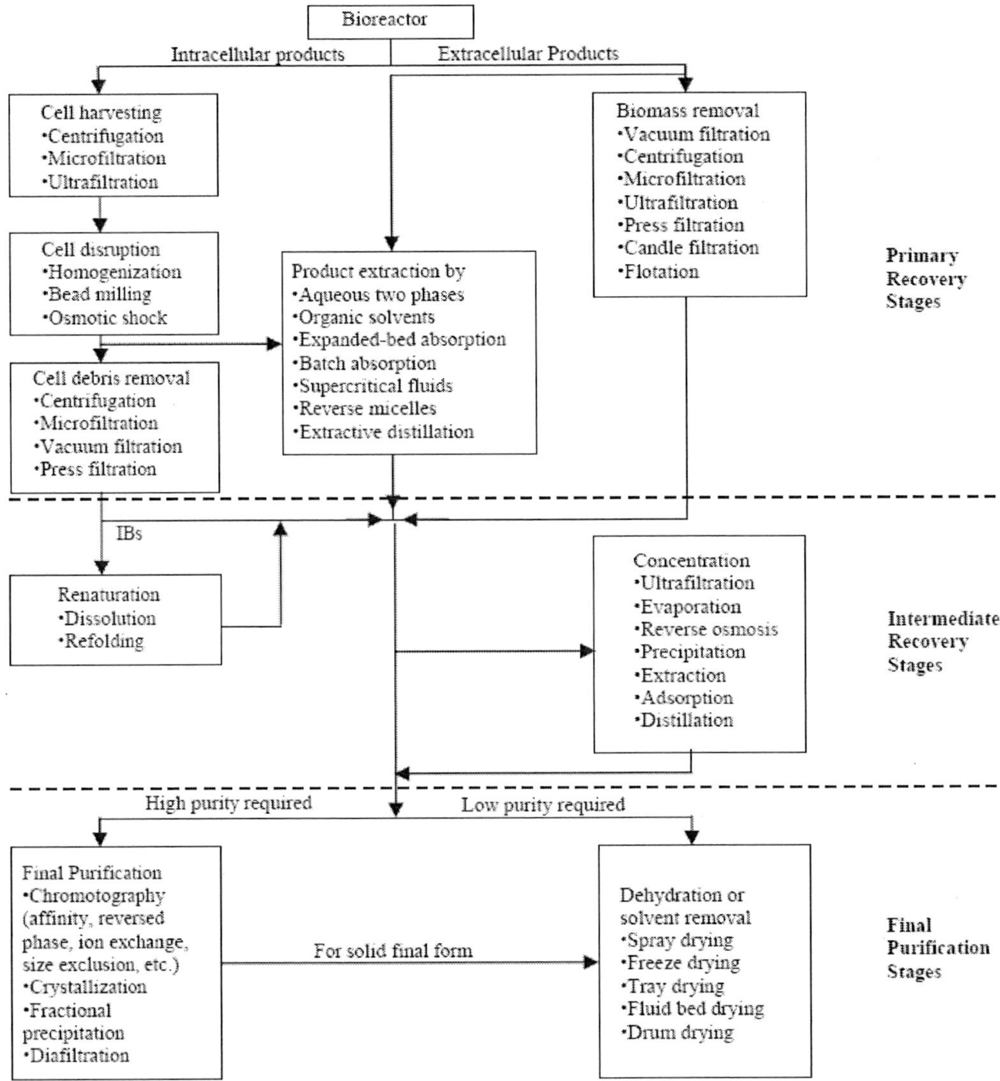

Figure 5. Overview of bioseparations [2].

2. State of the Science

Among separation techniques, mechanical filtrations are especially important in bioprocessing. These include micro-, ultra-, and nanofiltration; deep bed filtration; static and dynamic cross-flow filtration; electrofiltration; and centrifugation, involving both filtration and sedimentation. In these filtrations, fouling remains a major technical obstacle. In addition, the chemical and thermal separation techniques of ion exchange, chromatography, electrophoresis, crystallization, and extraction are also commonly used in bioprocessing. Research is currently underway in improving many bioseparation techniques, offering hope that major improvements in the efficiency of bioengineered processes can be achieved by incorporating new bioseparation technologies into process design, and several recent, excellent compilations of modern techniques are now available [3-5].

2.1. Physical Separations

Emerging physical bioseparation techniques include aqueous two-phase extraction, reverse micellar extraction, cloud point extraction, and magnetic and electrophoretic separation [6, 7].

2.1.1. Two-phase partitioning bioreactors. Among emerging techniques, the two-phase partitioning bioreactor appears to have great potential in enhancing the productivity of many bioprocesses. The approach integrates fermentation with a primary product recovery step by incorporating both organic and aqueous phases simultaneously, such that microbial growth occurs in the aqueous media and substrates and products partition into the organic phase based on their affinities for it. This approach allows controlled substrate delivery to the fermentation broth and effectively lowers product concentrations in the fermentation broth as well, promoting microbial health and activity. Although it is already practiced commercially, its effectiveness could still be improved by the discovery of low-cost solvents that are non-toxic for microbial growth [8, 9].

2.1.2. Non-solvent-based processing. A primary purpose of nonpolar organic solvents in bioprocessing is the dissolution of cell membranes to release intracellular products However, organic solvents are undesirable for several reasons: they are frequently derived from non-renewable resources such as petroleum; they are frequently toxic and/or carcinogenic, and consequently expensive to treat in disposal; and they frequently form non-aqueous phases that hinder biodegradation. As a result, efforts to develop efficient, non-solvent-based approaches to cell disruption and product purification deserve high priority.

The purification of polyhydroxyalkanoate (PHA) polymers provides an excellent example of the benefits possible with such approaches. Following fermentation, PHA-containing cells are separated from culture media by centrifugation, filtration, and/or flocculation, and cells are then disrupted to recover the polymer. Subsequent recovery, unfortunately, typically involves extraction of the polymer from biomass with large amounts of toxic and inflammable organic solvents (e.g., chloroform, methylene chloride, propylene carbonate, or dichloroethane). Another established method, again involving environmentally undesirable reagents, uses sodium hypochlorite for the digestion of non-PHA cellular materials, carrying the additional disadvantage that it partially degrades the PHA [10].

To improve the environmental friendliness of PHA processing, a non-solvent-based process has been developed by Zeneca Agrochemicals (now part of Syngenta; www.syngenta.com) to assist in the commercial production of poly[(R)-3-hydroxybutyrate] [poly(3HB)] and poly(3HB-co-3V) by Alcaligenes eutrophus. In this process, cells are brought to 80°C and treated with a mixture of hydrolytic enzymes, including lysozyme, phospholipase, and lecithinase that hydrolyze cellular components without degrading the polymer. The polymer can then be recovered as a white powder after washing and drying, demonstrating the technical feasibility of performing novel separations utilizing advanced bioengineering techniques. Cost of necessary reagents, such as the enzymes involved in this case, remains an important consideration, however, and must be addressed to enable such techniques to compete with existing routes to synthesis of plastics based on petrochemical sources [11]. Additional non-solvent based methods of cell lysis, including modern developments in enzymatic, chemical, and mechanical cell disruption, are considered in detail in the recent book, Bioseparations Science and Engineering [5].

2.2. Mechanical Separations

A great variety of membrane-based techniques are not only under development, but are already central to commercial integrated bioprocesses, enabling further avoidance of environmentally deleterious reagents in bioproces sing. These approaches separate products from culture media based on hydrophobicity, volatility, and/or affinity for membrane components, and offer the potential for great selectivity as well as low-energy and waste-disposal requirements in biorefining [12].

2.2.1. Pervaporation. Pervaporation is the separation of various liquid mixtures by partial vaporization through a non-porous membrane. The membrane acts as a selective barrier between the two phases: the liquid feed and the vapor permeate phases. It allows the desired component(s) of the liquid feed to dissolve within it and then to transfer through it by vaporization, resulting in a separation based primarily on differences in polarity rather than volatility. Pervaporation has shown great promise in separating alcohols such as ethanol and butanol from bioreactor media, as in the production of ethanol from rice straw by Pichia stipitis [13] as well as other azeotropic and close-boiling mixtures, including isomers [14].

A challenge associated with the use of pervaporation is the accumulation of less volatile components (higher alcohols) in fermentation media that can cause microbial growth inhibition [9]. Nevertheless, industrial uses for this method are widespread, and membrane technologies as well as integration of pervaporation with other unit operations are still showing promising improvement [14].

2.2.2. Size-based separations. Membranes have traditionally been used for size- based separations that require high throughput but relatively low resolution. Examples include microfiltration (MF) for clarification and sterilization as well as ultrafiltration (UF) for product concentration and buffer exchange Microfiltration is competitive with depth filtration, centrifugation, and expanded-bed chromatography for the initial harvest of products from bacterial, yeast, and mammalian cell cultures [15, 16, 5]. Use of 0.2 μm-rated MF membranes can yield a particle-free harvest solution requiring no additional clarification prior to purification, while larger pore-size membranes can be used to improve product yield and throughput when the filtrate is subsequently treated with a normal flow filter for final clarification. MF systems are generally operated at constant flux instead of constant transmembrane pressure to improve product yield and throughput, although this practice tends to exacerbate fouling problems [17]. An emerging solution to this problem is high-frequency, back-pulsing air scrubbing to clean the surface of the membrane continuously [18, 5].

Ultrafiltration, in turn, has become the method of choice for protein concentration, replacing size-exclusion chromatography in this application [19]. Plasmid DNA [20] and virus- like particles [21] can also be purified using UF. UF membranes are typically composed of polysulfone, polyethersulfone, or regenerated cellulose, and experience fouling problems analogous to those that plague MF membranes. Newly-developed composite cellulose membranes are less susceptible to fouling than are many synthetic polymers, and are more easily cleaned, but still possess excellent mechanical strength, thus outperforming other membrane materials. Nevertheless, cellulose has had a much longer development history than have synthetic polymers in membrane applications, and promising developments in the latter are still expected [22].

2.2.3. Membrane chromatography. Recent studies are generating renewed interest in membrane chromatography [23]. Diffusion limitations are less pronounced in membranes than in conventional bead packings, thus providing, in principle, a binding capacity independent of flow rate. Although the high, internal surface area provided by small pores in membranes is often offset by the reduction in convective flow, and it is challenging to achieve binding capacities competitive with bead packings, research into optimization of membrane chromatography continues [24, 25]. Especially encouraging are manifold designs with high membrane density and low retention volumes that can accomplish up to 100-fold concentrations in a single stage [5].

Flow distribution within membrane modules is an important consideration for optimal performance, especially in manifold designs, and careful design of entrance and exit regions is essential to provide even, well-distributed flow. In addition, sensing and maintenance of constant concentration of the retained species at the membrane surface is becoming a new robust control strategy, enhancing product yield and providing greater operational robustness with respect to variations in feed quality [26]. Finally, additional efforts are directed toward improving binding capacity, membrane selectivity, flow distribution, and flow rate, through adjustments in pore size, membrane chemistry, and membrane morphology [27, 28].

2.3. Microbe Engineering

Sometimes bioprocessing procedures can be simplified greatly by engineering an organism itself to produce a more convenient product, as in the development of DuPont's bio-based 1,3-propanediol (3G) process [29]. In nature, two separate microorganisms are required to convert glucose to glycerol and then to convert glycerol to 3G. To avoid the difficulty of purifying the glycerol, a genetically-engineered microorganism was developed that converted glucose directly to 3G. With increasing advances in metabolic pathway engineering (see section II.A.2.7), such avenues are expected to become increasingly available.

3. Research Priorities

Separation technologies to facilitate commercial success of biomass conversions include those suitable for low-molecular-weight organic acids, organic esters, diacids, and alcohols; gases such as H2; and biobased oils such as biodiesel and biolubricants. Among these, advances in membrane technologies and in processes utilizing environmentally-benign solvents promise especially great benefits.

3.1. Membrane Techniques

The development of new and/or improved membrane materials that provide increased selectivity and specificity for the desired substances, as well as increased flux with stability and robustness, is of central importance to the membrane-based techniques discussed below [5].

3.1.1. Pervaporation. The use of pervaporation to remove either water or bioproducts from bioreactor media appears promising. Continued support for new membrane materials, new module and process designs, and improved theoretical understanding and modeling of the pervaporation process should therefore be pursued. The work of Vane and colleagues at the EPA National Risk Management Research Laboratory (NRMRL) is a noteworthy example of efforts in the development of pervaporation modeling and performance prediction software [30].

3.1.2. Micro- and ultrafiltration. Microfiltration and ultrafiltration promise to become major unit operations in the emerging biorefinery arena. The development of new materials for UF and MF, including porous metals and ceramics as well as polymers, is therefore an important priority. Similarly, nanofiltration and reverse osmosis are becoming increasingly important, with recent developments in nanotechnology promising to yield new materials with significantly improved fluxes and selectivities [5].

3.1.3. Membrane chromatography. An improved understanding of the interactions between culture media components and synthetic polymers suitable for membranes would greatly facilitate the design of synthetic substrates for use in membrane chromatography. Among those, ligand-binding and sterically-interacting species should be investigated closely to improve the selectivity of membrane chromatography while maintaining acceptably high throughput

3.1.4. Antifouling techniques. Fouling is a persistent problem among membrane technologies, with the result that methods to diminish fouling of membranes and ion exchange materials, as well as to remove impurities such as salts or acids that cause complications in downstream processes, are high priorities in the advancement of bioseparations.

3.2. Environmentally Benign Solvents

New renewable, biodegradable solvents are needed to support environmentally-friendly extraction processes. Supercritical CO_2, a highly compressed phase of CO_2 possessing properties of both liquid and gas phases, is one benign solvent that has already achieved great popularity and that has the potential to contribute performance, cost-effectiveness, and sustainability to separations of both biofuels and biomaterials [31].

3.3. Integrated Modules

Combined- or hybrid-unit operations in which a bioreactor is integrated with a bioseparation module, as in two-phase reactor systems, are particularly attractive as means to overcome limitations inherent to bioprocessing. These are particularly desirable for their potential to remove products as they are synthesized, alleviating the nearly universal problem of product inhibition in culture media [5].

4. Commercialization

One of the primary limitations to commercialization of biobased products is the high cost of isolating the desired products. Accordingly, development of advanced separation technologies should be an integral part of a comprehensive program for reducing costs and encouraging commercialization. Affinity-based separations, environmentally-benign extractions, and membrane separations hold the promise of operating at low temperatures and thereby reducing the demand for expensive energy presented by thermal separation processes such as distillation. Currently, fouling, other flux limitations, and sub-optimal selectivities limit the larger scale deployment of the more sustainable separation techniques, with the implication that research should focus on these areas.

REFERENCES

[1]   Perry, R. H., and D. W. Green, Eds. (1997). *Perry's Chemical Engineers' Handbook.* McGraw Hill, New York.

[2]   Petrides, D. P., C. L. Cooney, and L. B. Evans (1989). *An introduction to biochemical process design,* in Chemical Engineering Problems in Biotechnology (M. L. Shuler, Ed.). American Institute of Chemical Engineers, New York, pp 351-391.

[3]   Keller, K., T. Friedmann, and A. Boxman (2001). *The bioseparation needs for tomorrow,* Trends Biotechnol 19:438-441.

[4]   Ladisch, M. R. (2001). *Bioseparations Engineering: Principles, Practice, and Economics.* Wiley-Interscience.

[5]   Todd, P., S. R. Rudge, D. P. Petrides, and R. G. Harrison, Eds. (2003). *Bioseparations Science and Engineering.* Oxford University Press.

[6]   Karumanchi, R. S. M. S., M. Dueser, and P. W. Todd (2000). *Multistage magnetic and electrophoretic extraction of cells, particles and macromolecules,* Adv Biochem Eng Biotechnol 68:143-190.

[7]   Karumanchi, R. S. M. S., S. N. Doddamane, C. Sampangi, and P. W. Todd (2002). *Field-assisted extraction of cells, particles and macromolecules,* Trends Biotechnol 20:72-78.

[8]   Gu, T. (2000). *Liquid-liquid partitioning methods for bioseparations,* in Handbook of Bioseparations (S. Ahuja, Ed.). Academic Press, San Diego, pp 329-364.

[9]   Malinowski, J. (2001). *Two-phase partitioning bioreactors in fermentation technology,* Biotechnol Adv 19:525-53 8.

[10] Williamson, D. R., and J. F. Wilkinson (1958). *The isolation and estimation of the poly-beta-hydroxybutyrate inclusions of Bacillus species,* J Gen Microbiol 19:198-209.

[11] Byrom, D. (1987). *Polymer synthesis by micro-organisms: technology and economics,* Trends Biotechnol 5:246-250.

[12] Bailly, M., H. Balmanna, P. Aimar, F. Lutin, and M. Cheryan (2001). *Production processes of fermented organic acids targeted around membrane operations: design of the concentration step by conventional electrodialysis,* J Membr Sci 191:129.

[13] Nakamura, Y., T. Sawada, and E. Inoue (2001). *Mathematical model for ethanol production from mixed sugars by* Pichia stipitis, J Chem Technol Biotechnol 76:586-592.

[14] Smitha, B., D. Suhanya, S. Sridhar, and M. Ramakrishna (2004). *Separation of organic- organic mixtures by pervaporation: a review*, J Membr Sci 241:1-21.

[15] van Reis, R., L. C. Leonard, H. C. Chung, and S. E. Builder (1991). *Industrial scale harvest of proteins from mammalian cell culture by tangential flow filtration*, Biotechnol Bioeng 3 8:413-422.

[16] Dorgan, J. R. (1992). *Polymer Membranes for Bioseparations*, in Polymer Applications for Biotechnology (D. S. Soane, Ed.). Prentice Hall, Englewood Cliffs, pp 314.

[17] Belfort, G., R. H. Davis, and A. L. Zydney (1994). *The behavior of suspensions and macromolecular solutions in crossflow microfiltration*, J Membr Sci 96:1-58.

[18] Levesley, J. A., and M. Hoare (1999). *The effect of high frequency backflushing on the microfiltration of yeast homogenate suspensions for the recovery of soluable proteins*, J Membr Sci 158:29-39.

[19] Kurnik, R. T., A. W. Vu, G. S. Blank, A. R. Burton, D. Smith, A. M. Athalye, and R. van Reis (1995). *Buffer exchange using size exclusion chromatography, countercurrent dialysis, and tangential flow filtration: models, development, and industrial application*, Biotechnol Bioeng 45:149-157.

[20] Kahn, D. W., M. D. Butler, G. M. Cohen, J. W. Kahn, and M. E. Winkler (2000). *Purification of plasmid DNA by tangential flow filtration*, Biotechnol Bioeng 69:101-106.

[21] Cruz, P. E., C. C. Peixoto, K. Devos, J. L. Moreira, E. Saman, and M. J. T. Carrondo (2000). *Characterization and downstream processing of HIV-1 core and virus-like-particles produced in serum free medium*, Enzyme Microb Technol 26:2661-2670.

[22] van Reis, R., and A. L. Zydney (1999). *Protein ultrafiltration, in Encyclopedia of Bioprocess Technology: Fermentation, Biocatalysis, and Bioseparation* (M. C. Flickinger and S. W. Drew, Eds.). John Wiley & Sons, New York, pp 2197-2214.

[23] Zeng, X., and E. Ruckenstein (2000). *Membrane chromatography: preparation and applications to protein separation*, Biotechnol Prog 15:1003-1019.

[24] Schweitzer, P. A. (1997). *Handbook of Separation Techniques for Chemical Engineers*. McGraw-Hill.

[25] Noble, R. D., and P. A. Terry (2004). *Principles of Chemical Separations with Environmental Applications*. Cambridge University Press.

[26] van Reis, R., E. M. Goodrich, C. L. Yson, L. N. Frautschy, R. Whiteley, and Z. Al (1997). *Constant Cwall ultrafi ltration process control*, J Membr Sci 130:123-140.

[27] Sarfert, F. T., and M. R. Etzel (1997). *Mass transfer limitations in protein separations using ion-exchange membranes*, J Chromatogr A 764:3-20.

[28] Klein, E. (2000). Affinity membranes: a 10 year review, J Membr Sci 179:1-27.

[29] Laffend, L. A. (1997). *Bioconversion of a Fermentable Carbon Source to 1,3-propanediol by a Single Microorganism*, E.I. DuPont de Nemours and Company, Patent 5686276.

[30] Pend, M., S. X. Liu, and L. M. Vane (2003). *Profiling concentration gradient in a tubular membrane pervaporation module: a modeling approach*, Intl Comm Heat Mass Transfer 30:755-764.

[31] Kemmere, M. F., and T. Meyer, Eds. (2005). *Supercritical Carbon Dioxide in Polymer Reaction Engineering*. John Wiley & Sons.

Chapter III

BIOMATERIALS

A. PROBLEM AND SIGNIFICANCE

1. Overview of Importance

The worldwide production of plastics reached 260 billion pounds per year at the end of the 20th century, with a value of over \$310 billion to U.S. economy in 2002 [1]. Large quantities of petroleum are used to produce present-generation plastics, but oil is of finite supply, and as world economies continue to develop oil it will become more and more expensive [2]. Additionally, pollution results from the manufacture, use, and disposal of plastic materials. As the world's finite supplies of petroleum are used up, and as the growing industrial economies of China and other regions continue to rapidly boost demand, oil prices are skyrocketing, along with the prices for all products that rely on oil supplies.

Moreover, the increased demand is driving us to drill in sensitive areas and to use lower-grade crude oils that are less economical and that contain contaminants that threaten the environment. However, plastics offer profound societal benefits, including increased agricultural production, reduced food spoilage, reduced fuel consumption in lighter-weight vehicles, better health care, and low-cost, net-shape manufacturing. We need plastics, but the financial hardship consumers are feeling at the gas pump from increasing prices is also sharply impacting the plastics industries where production costs are rising and being passed on to the consumer. What will happen to our environment, to human and animal health, and to the plastics industries–the fourth largest manufacturing sector of the U.S. economy employing more than 1.2 million citizens (1)–if sustainable technologies are not developed and deployed? While energy recovery through combustion, recycling, and minimizing use of plastics all aid pollution prevention, a new paradigm is emerging that holds great promise— namely the production of plastic materials using renewable resources as feedstocks. This approach is becoming increasingly viable due to advances in industrial biotechnology.

It can be argued that using bioengineering techniques to make plastic materials is more economically favorable than using these same techniques to make fuels. The reason for this economic advantage is simply that plastics have a greater value than fuels. Consider gasoline at a retail price of \$3 per gallon; the price before taxes would be approximately \$2.25–\$2.65 per gallon. This means that for approximately \$2.50, about 7 pounds of gasoline can be purchased; that is, the price per pound is about \$0.35. For comparable crude oil prices, commodity plastic resins would cost anywhere from \$0.50 per pound to upwards of \$3.00 per pound. As discussed further below, biobased plastics are already economically competitive with conventional petroleum based plastics because of the higher price they can command when they have specific materials properties. It can be argued that industrial bioengineering for commodity production will first be utilized on the most economic targets and that plastics are more economically attractive than fuels.

Pursuing economically viable plastics using bioengineering techniques can also affect fuel prices to a substantial degree. This economic impact is due to several factors. Presently about 65 percent of petroleum goes to transportation fuels, [3] however, if the 5–10 percent of petroleum going into plastic materials can be substituted, the marginal cost of petroleum could decrease significantly. This is due to the fact that market surpluses and shortages are, in fact, relatively small perturbations on a large base number. Perhaps more significantly, producing bioplastic materials provides an opportunity for economic integration. That is, by having integrated biorefineries in which high-margin plastics are coproduced with biofuels, economic production of the lower-margin fuel products can be produced.

There are a number of materials that can be made that start with renewable resources and use bioengineering techniques. These include thermoplastics, thermosets, foams, pressure sensitive adhesives, various biocomposites, and coatings. This document focuses on the emerging material classes where bioengineering techniques play the largest role; that is, where engineered organisms play a prominent role. Only a brief synopsis of materials available from bioresources using more traditional chemical routes is given; several excellent monographs are available that discuss biobased plastics and composites [4-6].

2. Conventional Plastics

Considerable environmental pollution occurs as a result of the production, use, and disposal of conventional plastics. The absolute mass of plastics produced in a given year is small compared to that of fuels; however, due to their persistence they present unique challenges when released into the environment. Current polymer materials are nearly all derived from petrochemical sources, contributing significantly to greenhouse gas emissions during both production and incineration [7, 8]. Because producing plastics from renewable resources provides a real economic opportunity, perhaps the best possible substitution strategy would be to make non-degradable, persistent plastics from renewable resources; this strategy could provide at least a relative sequestration of CO_2 from the atmosphere, thereby mitigating effects of global climate change [9].

3. Plastics from Renewable Resources

Through application of bioengineering in conjunction with traditional chemical processing techniques, bioplastics production is proving viable in a number of commercially produced materials. Most notable are the production of PLA by Cargill-Dow under the trade name of Natureworks[TM] (http://www.packexpo.com/ve/82489/main.html); the production of Sorona[TM] polyesters, which contain a 1,3 -propanediol monomer derived from renewable feedstocks, by DuPont (http://www.dupont.com/sorona/backgroundsoronapolymer.html); and the production of PHA by Metabolix, Inc. (www.metabolix.com).

Other environmentally benign plastic materials include those available directly from plants, such as starches, starch-protein composites, triglycerides, and cellulosics (www.greenplastics.com). It is important to understand that many such plant-based plastics are presently competitive with petroleum-based materials on a cost-performance basis.

A clear example of this competitiveness is the case of Nylon-11 (polyamide-1 1), which is made from castor oil extracted from castor beans. This tough and resilient material is marketed as Rilsan by the Arkema Company (www.arkema-inc.com) and is widely used in powder-coating applications. These demanding applications include protective coverings for submerged pipelines and other industrial piping. In these uses, it is the superior properties of the plastic, rather than its renewable nature, which has developed the market. While biodegradable plastics are also available from petroleum-derived chemicals, this is not accomplished using bioengineering techniques and so these materials fall outside the scope of this document.

Polymer Nomenclature

Because the challenges associated with bioplastics development depend on the detailed chemistry and physics peculiar to polymers, a set of useful terms is presented here for readers less familiar with this specialized topic.

Polymer. A high-molecular-weight organic compound with a repeating unit (a monomer) that constitutes its structure. While typically represented by a long linear structure, different chain architectures are possible by including branching points and other monomers. If two or more monomers are present in the molecule it is referred to as a copolymer. Block copolymers consist of long strings of the same monomer connected to other long stretches of a different monomer whereas random copolymers have a random sequential arrangement of the different monomers. Branched polymers consist of linear sections that diverge at a point so that three or more strands emerge from the common branch point. In hyperbranched polymers the divergence from one branch point leads to another and each of those branches leads to another branch point and so on to create a highly branched or arborescent (tree like branching) polymer.

Glass Transition Temperature (Tg). The temperature at which the reversible change in an amorphous polymer or in the amorphous regions of a semi-crystalline polymer change to (or from) a hard and brittle glassy material to a soft, viscous, rubbery material. Hardness, thermal expansibility, and specific heat all change abruptly at Tg, however, it is not a true thermodynamic transition as it shows a dependence on cooling rate.

Melting Temperature (Tm). The temperature where all crystalline structure is lost to yield a liquid. Scientifically, it is incorrect to talk about the melting temperature of a fully amorphous polymer, however, in practice a melt-flow temperature of 50 °C above the Tg is sometimes used.

Polydispersity Index (PDI). The ratio of weight averaged to number averaged molecular weight. This index measures the breadth of the molecular weight distribution in a polymer sample. If all polymer chains in a sample had exactly the same length, the PDI would be 1.0; typical polymer samples have values from 1.5 to 30.0.

Tacticity. Refers to the geometric arrangement of substituent groups (e.g., a methyl group) along a polymer chain backbone. Syndiotactic indicates the substituents alternate regularly on opposite sides, isotactic means the substituents are all on the same side, and atactic means the order is random.

4. Renewable Polymer Production

Conceptually, three tiers of polymer production from renewable resources exist, as shown in Figure 6 [10]. Polylactic acids (PLAs) and Sorona polymers are produced in a three-stage process. The feedstock is first harvested in the form of crops, crop residuals, or other biomass, and refined to yield plant sugars (primarily glucose and xylose). A secondary stage of production consists of forming the subunit monomers via fermentation and separation. Finally, the third stage consists of traditional chemical processing to form the polymers that will ultimately be molded into plastic parts.

In a two-stage production system, plant sugars are used as substrates to support growth of bacteria that synthesize polymeric materials directly to store their own excess energy, much as an animal synthesizes glycogen. Such materials are being developed by a number of groups and corporations around the globe, including Proctor and Gamble (under the trade name Nodax, www.nodax.com) and Metabolix, Inc. (www.metabolix.com)—both of which now claim the ability to provide commercial scale samples. It can also be argued that the many interesting and useful materials being made from soybean and other plant oils [6] fall into the two-stage production scheme, but with the second stage consisting of chemical rather than biochemical transformation.

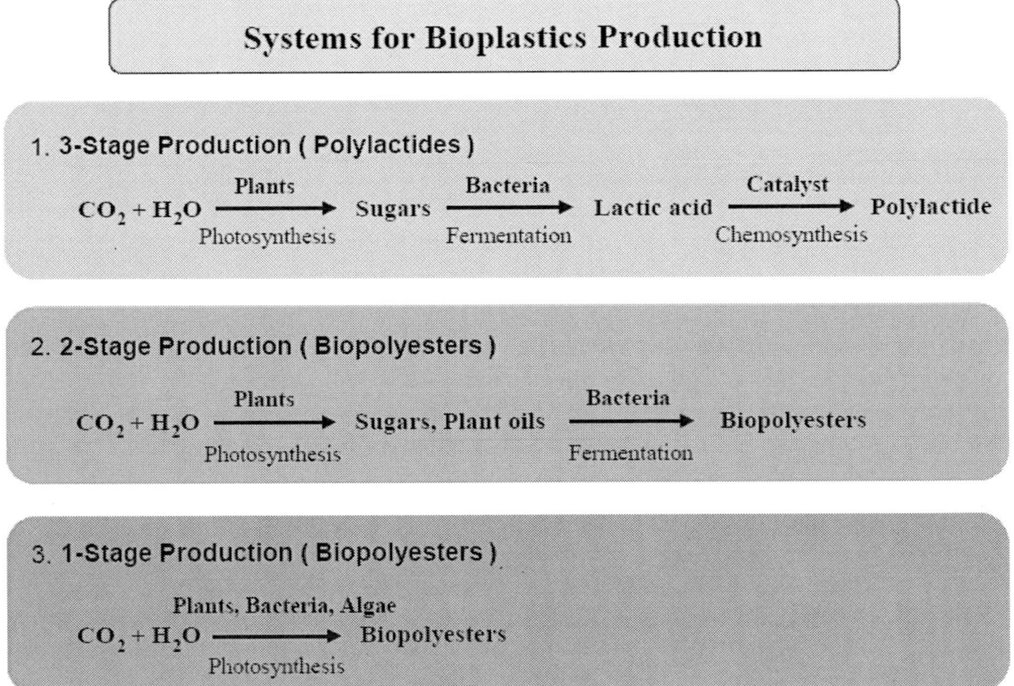

Figure 6. Systems for the production of materials by bioengineering [10].

In single-stage production, the material of interest is grown directly within the plant. This technology is the most futuristic of the three but holds the greatest promise for additional environmental benefits, particularly the direct capture of atmospheric CO^2 within the plastic; it is now being pursued energetically by Metabolix, Inc. for production of PHA within both the genetic model organism, Arabidopsis, and within the hardier switchgrass. While technical issues regarding both genetic engineering of the plants and separation of product polymers from plant tissues persist, ongoing research efforts have provided steady progress in these areas [11, 12].

5. Environmental Benefits of Green Plastics

Since 1993, the International Organization for Standardization (ISO) has been developing LCA programs to analyze material and energy requirements of various products [13]. Life-cycle Impact Assessment (LCIA), in turn, is the part of LCA in which the inventory of a product's material flows is translated into environmental impacts and resource consumption [14].

Although the LCIA of plastic products is still in the early stages of development, it will soon be possible to compare the environmental impacts of various green plastics with one another and with the conventional polyolefins that make up more than 90 percent of current plastics production in a quantitative, reliable way.

Controversy does exist regarding the extent to which the production of bioplastics serves the principles of environmental sustainability. Gerngross and Slater, for example, have argued that the conversion of biomass to biopolymers is energy-intensive (involving fertilizer production, farming, corn wet-milling, fermentation, and polymer purification, for example) and results in significantly greater net CO2 emissions than the synthesis of petroleum-based plastics [9]. While conceding that renewable energy sources (e.g., solar, wind, geothermal, etc.) could be used to improve the environmental profiles of these processes, they argue that greater environmental benefit could be obtained by using that energy to displace fossil fuels in other applications. Gerngross and Slater also question the benefits of biodegradability, with the rationale that non-degrading polymers produced from renewable resources could be used to sequester atmospheric CO_2.

The gaseous emission resulting from biodegradation of any biopolymers also represents an important consideration, although few agree that the alternative, accumulation of plastic debris, is desirable. In landfills, oxygen is limited, and organic matter is often degraded anaerobically to yield a mixture of CO_2 and CH_4 rather than pure CO_2. Because CH_4 is a 23-fold more potent greenhouse gas than CO_2 [15], anaerobic degradation of bioplastics is potentially quite deleterious to the environment. In a composting environment, however, in which regular mixing ensures aerobic degradation, only CO_2 is released, causing a process supported by sustainable agriculture and composting to be effectively CO_2-neutral [16]. In addition, another promising disposal option for bioplastics may be waste incineration with recovery of the energy generated, which then can be used to displace fossil fuel-derived energy [17].

Gruber has noted that a life-cycle assessment performed by Gerngross considered only the production of PHAs by microbial fermentation, and that other materials such as PLA hold the promise of lower energy requirements for production. This perspective has, in fact, been supported by other more recent life-cycle analyses [18]. A review of 20 such LCA studies of biodegradable polymers indicates that starch, the major component of approximately 75 percent of green plastics production, offers important environmental benefits compared to conventional polymers [19]. Concerning commercially produced biodegradable polymers, the environmental benefits of PLA, currently accounting for 10–15 percent of production, and of biodegradable polymers made from nonrenewable resources, accounting for approximately 10 percent of production, seem to be smaller than starch-based thermoplastics, but still greater than conventional polymers. For microbial polyesters, which currently make up a very small part of total green plastics production, the environmental advantage seems to be small (or perhaps nonexistent), but the fermentation technologies for producing them are among the most recently developed, and both the production method and the scale of production can influence evaluations of the overall environmental balance [20]. Additionally, the prospect of producing these plastics in transgenic plants completely alters the environmental consequences of production, thereby opening up improved life-cycle possibilities [12, 21].

Given the inherent problems associated with persistent plastics in the environment–increasing pressure on landfill space, concerns over climate change, and the economic reality that biobased plastics are already competing in the market without subsidies–applying the tools of industrial biotechnology to the production of environmentally benign plastics is a particularly vibrant area of scientific and commercial activity.

REFERENCES

[1]   Society of the Plastic Industries (SPI) (2003). *Size and Impact of the U.S. Plastics Industry. Executive Summary,* http://www.plasticsindustry.org/industry/impact.htm.

[2]   Dreffeyes, K. S. (2001). Hubbert's Peak: The Impending World Oil Shortage. Princeton University Press, Princeton, NJ.

[3]   Board of Transportation Statistics (2004). *Overview of U.S. Petroleum Production, Imports, Exports, and Consumption,* Washington, D.C., http://www.bts.gov/ publications/national_transportation_statistics/2004/html/table_04_0 1 .html.

[4]   Kaplan, D. L. (1998). *Biopolymers from Renewable Resources.* Springer-Verlag, Berlin.

[5]   Stevens, E. S. (2002*). Green plastics: An Introduction to the New Science of Biodegradable Plastics.* Princeton University Press, Princeton, NJ.

[6]   Wool, R. P., and X.S. Sun (2005). *Bio-Based Polymers and Composites.* Elsevier Academic Press, Burlington, MA.

[7]   Harper, C. A. (2000). *Modern Plastics Handbook.* McGraw-Hill Professional, New York.

[8]   Strong, A. B. (1999). *Plastics: Materials and Processing,* 2nd Edition. Prentice Hall, New York.

[9]   Gerngross, T. U., and S.C. Slater (2003). *Biopolymers and the environment,* Science 299:822-825.

[10]   Doi, Y. (2004). *In Life Cycle Assessment of Microbial Polyesters, BioEnvironmental Polymer* Society Meeting, Denver.

[11]   Skraly, F. (2002). *Bioplastics, in Encyclopedia of Environmental Microbiology.* John Wiley & Sons, New York.

[12]   Snell, K. D., and O.P. Peoples (2002). *Polyhydroxyalkanoate polymers and their production in transgenic plants,* Metab Eng 4:29-40.

[13] International Organization for Standardization Environmental Management (1997*). Life Cycle Assessment—Principles and Framework.* Geneva, Switzerland, ISO 14040.

[14] Hauschild, M. (2005). *Assessing environmental impacts in a life-cycle perspective,* Environ Sci Technol. 39:81A-88A.

[15] U.S. Department of Energy (2003). *U.S. Climate Change Technology Program: Research and Current Activities,* Washington, D.C., http://www.climatetechnology.gov/library/2003/currentactivities/car24nov03 .pdf DOE/PI-0001.

[16] Schlesinger, W. H. (1997). *Biogeochemistry: An Analysis of Global Change.* Academic Press, San Diego, CA.

[17] Holmgren, K., and D. Henning (2004). *Comparison between material and energy recovery of municipal waste from an energy perspective: a study of two Swedish municipalities,* Res Cons Recyc 43:51-73.

[18] Vink, E. T. H., K. R. Ra´ bago, D. A. Glassner, and P.R. Gruber (2003). *Applications of life cycle assessment to Nature Works polylactide (PLA) production,* Polymer Degrad and Stab 80:403–419.

[19] Patel, M. (2002). *Do Biodegradable Plastics Fulfill the Expectations Concerning Environmental Benefits? 7th World Conference on Biodegradable Polymers & Plastics,* Tirrenia, Italy.

[20] Stevens, E. S. (2005). *Green plastics: Environment-friendly technology for a sustainable future,* http://www.greenplastics.com/.

[21] Martin, D. P., O. P. Peoples, and S. F. Williams (2000). *Methods for Isolating Polyhydroxyalkanoates from Plants.* Patent US 6,709,848.

B. POLYLACTIDES

1. Introduction

The family of bioplastics known as PLA encompasses the set of polymers of lactide, a cyclic dimer produced by the dehydration of lactic acid, which represents a highly promising and versatile category of biomaterials. The development of PLA into a commodity polymer has spanned over six decades of research and design from inception to the present commercial utility. The recorded history of PLA development began in 1932 when Carothers and colleagues documented the earliest attempted polymerization and depolymerization of oligomeric lactides in the Journal of the American Chemical Society [1]. In 1954, the DuPont Corporation synthesized high molecular-weight PLA with improved lactide purification techniques, as well as antimony trioxide and antimony trihalide as polymerization catalysts [2]. Later, methods were developed to produce high molecular weight PLA with properties sufficient for competition with traditional oil-derived polymers. Higher molecular-weight PLA was found to have interesting properties, but its production was prohibitively expensive. In the 1960s, several researchers investigated relationships between the chemical structures of lactide monomers and the configurations and crystalline structures of resulting PLAs [3-7]. Because of its inherent biodegradability, PLA was one of the earliest polymers used in biomedical applications; Kulkarni et al. demonstrated the human body's ability to absorb PLA-based sutures in 1966 [8]. Work in the 1970s and 1980s focused primarily on discovery of new catalyst systems for polymerization, improving characterization with new analytical techniques, and investigation of new medical applications [9].

Recently, the need for more biofriendly polymers has led to the study of additional catalyst systems, new types of copolymers, and polymerization mechanisms. In the early 1990s, methods were developed for the continuous production of both lactide and PLA. A major step in the commercialization of PLA occurred when Cargill Corporation developed its method to produce PLA in a continuous process [10-13]. A joint venture of Cargill with Dow Polymers (CDP) was then formed in 1997 to develop further the potential of PLA as a commodity polymer. A production facility opened in 2002 and was built in Blair, Nebraska with the capability of producing approximately 300 million pounds of PLA per year. Large-scale production of PLA has dramatically decreased the cost of PLA resins and is now enabling it to compete with established petroleum-based materials (Wall Street Journal, "One Word of Advice: Now it's Corn," October 12, 2004). Dow Chemical recently exited the joint venture in 2004 and the Blair facility is operated by a wholly owned subsidiary of Cargill called Natureworks (www.natureworksllc.com). This corporation is now producing and selling large quantities of plastics being used in products ranging from clothing to food packaging.

2. PLA Biosynthesis, Biodegradation, and Environmental Impact: Overview

PLA is presently commercially produced utilizing corn as the feedstock according to the process diagramed in Figure 7. Corn first undergoes the traditional milling process to produce unrefined glucose (dextrose), after which microorganisms ferment the glucose to lactic acid. Importantly, this fermentation process is anaerobic: a wide variety of literature suggests that anaerobic fermentations hold great advantage over aerobic fermentations when CO_2 balance is of concern (discussed below). After fermentation, the resulting lactic acid is formed into cyclic lactide through dimerization using reactive distillation. The lactide ring is then polymerized to produce PLA.

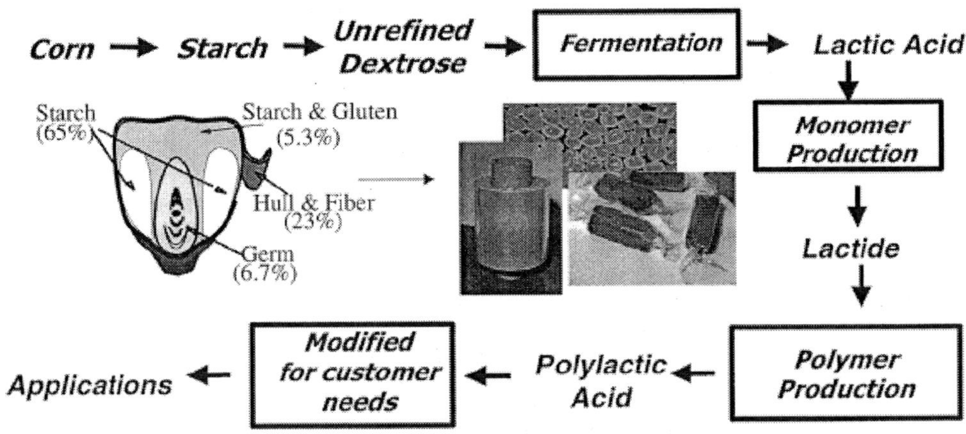

Figure 7. Commercial route to biobased PLA plastics.

The degradation of polymeric materials in the environment is a critical area of concern due to the large quantity of plastics generated. Each year, over 70 million tons of polymers are produced which end up in landfills [14, 15]. Breakdown of PLA in the environment can occur biotically or abiotically [16]. In the absence of sufficient microbial activity or oxygen, hydrolysis becomes the predominant pathway for degradation, while in aerated composting environments, biotic processes can degrade PLA rapidly (within weeks) and completely [17, 18]. Regardless of the process, degradation is also subject to the form of the plastic and purity of the PLA used [19]. As an alternative to biodegradation, waste PLA can be recycled into lactic acid, which can then be reformed into lactide and repolymerized [16, 17].

To assess the environmental impact of PLA synthesis and use, a comprehensive LCA enabling apples-to-apples comparisons with petrochemical-based thermoplastics was recently undertaken (Figure 8) [20]. Among the most notable benefits of PLA shown were reductions in both fossil-fuel use and global warming potential, even assuming use of fossil-based energy sources for agriculture and processing. Compared to most traditional hydrocarbon-based polymers, PLA uses 3 0–50 percent less fossil-fuel energy and results in lower CO2 emissions by 50–70 percent. Whereas conventional thermoplastic polymers require oil as their source of monomers and additional fossil fuels for processing, solar energy provides approximately one- third of the gross energy requirement in PLA production; in addition, processing of oil into conventional plastics releases even greater amounts of CO2 than does PLA production [20]. On October 11, 2005 Natureworks announced that they would purchase renewable energy certificates to offset the fossil energy being used in the production of PLA. By doing so, they claim to be producing the first-ever greenhouse gas-neutral commercial plastic (see www.natureworks.com).

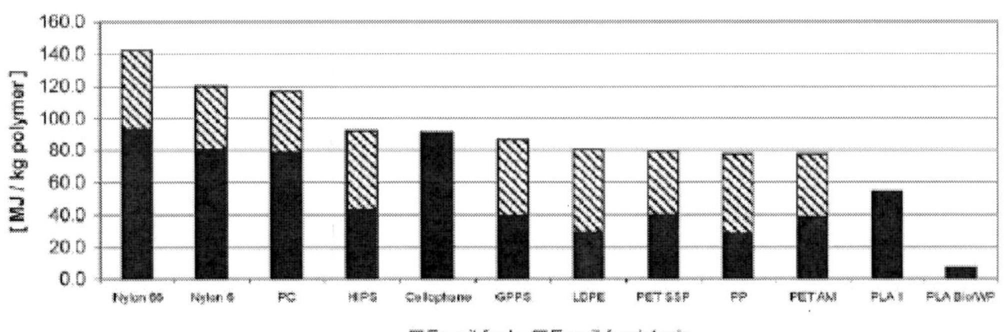

PLA1 represents present technology; PLA Bio/WP is the projection for the production of PLA from agricultural waste using wind power

Figure 8. Life cycle analysis results for energy content of various thermoplastic polymers [20].

3. State of the Science

3.1. Physical Properties

The bulk properties of PLA are greatly affected by the molecular weight of the polymer, the chain architecture (branched vs. linear), and the degree of crystallinity in the polymer [21, 22]. The amount of crystalline character within a type of PLA, in turn, is determined by the relative proportions of L- and D-lactide in the polymer backbone. A diagram of representative backbones is provided in Figure 9.

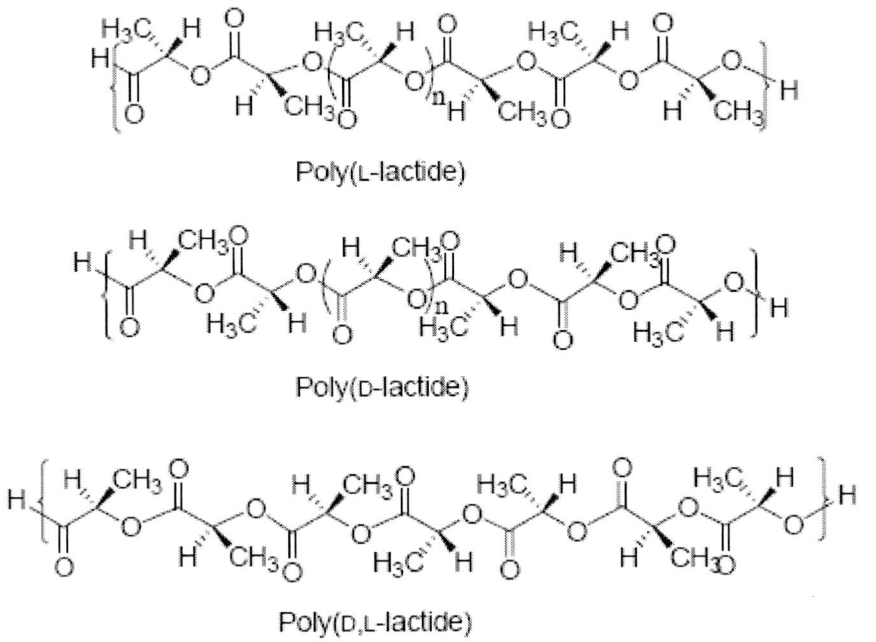

Poly(L-lactide)

Poly(D-lactide)

Poly(D,L-lactide)

Figure 9. Stereochemistry of L, D, and D,L-PLA backbones.

PLA samples containing 87.5 percent L-lactide are completely amorphous, while samples with 92 percent L-lactide possess some crystallinity [14]. Polymers from 100 percent L - lactide can be nearly half-crystalline [16]. The Tm range of crystalline PLA is 145-186°C [21], although a blend of 100 percent D-lactide and 100 percent L-lactide polymers in a stereo-complex form a closely packed crystalline structure that increases the T_m to 230°C [23]. The appearance of the PLA is also affected by the crystalline content. Amorphous PLA and low-crystalline PLA are clear materials with high gloss, while highly crystalline PLA is an opaque white material.

The molecular weight, structure, and crystallinity of PLA play important roles in its mechanical properties as well, including tensile strength, tensile modulus, and percent elongation to break. These are comparable to those of poly(ethylene terephthalate) (PET), PP, and polystyrene (PS), fortunately, presenting the possibility that PLA variations may replace thermoplastics in many applications. Investigation of such applications is an active area of research in Japan, North America, and Europe [17, 21, 22].

3.2. Applications

A wide range of products can be produced from PLA and PLA derivatives, including structural support and drug delivery in medical applications, tough fibers in woven and non-woven products, and in molded or extruded consumer products.

In initial medical applications, lactide and glycolide copolymers were used to make absorbable sutures capable of replacing denatured collagen or catgut [14]. Materials based on

PLA and PLA copolymers have also been designed to replace metal and other non-absorbable polymers as therapeutic aids in surgery, including pins [24], plates [25], screws [25, 26], suture anchors [27], and intravascular stents [9, 28]. The advantages of PLA and PLA glycolide copolymers for these applications comes from the ability of the body both to degrade the polymers and to metabolize the degradation products over time, leaving no residual foreign material in the body [8]. The ability to tune PLA degradation times is also relevant to the field of drug delivery, as drugs encapsulated in polymers can be released based on the known degradation time of PLA [29] or PLA copolymers [30, 31], even allowing the specific targeting of organs [32]. Other medical applications currently being pursued include dressings for burn victims [33], substrates for skin grafts [24], and dental applications [9]. Medical usage in devices, sutures and drug delivery systems, while technologically important, nevertheless represents only a small opportunity for displacing less environmentally benign plastics.

Fortunately, PLA also has properties suitable for commercial fiber products. Fibers can be produced from either pure liquid polymer, in a melt-spun process [34, 35], or from polymer dissolved in an organic solvent, in a solution-spun process [36, 37]. Advantageously, fibers produced from PLA have lower processing temperatures than PET fibers and therefore require lower energy input during processing. In applications, non-woven fibers are able to wick moisture without absorbing it and can therefore be used in products such as diapers. Materials produced from woven fibers have additional desirable properties, including favorable hand and touch, drape, wrinkle resistance, rapid wicking, and low-moisture absorbance. Woven PLA fiber-based materials are highly resilient to wear, have excellent UV resistance, and low inflammability. Applications for the PLA fibers or blends of PLA fibers with other natural fibers, such as silk, cotton, or wool, therefore include clothing, carpets, upholstery, and draperies [38].

The greatest opportunity for PLA to displace less benign materials is in the area of packaging for both food and consumer products. This opportunity arises because the properties of PLA allow improved function as well as diminished environmental impact. PLA can be used in traditional polymer processing operations such as injection molding, blow molding, extrusion, and extrusion coating. As a result, lids, trays, and clamshells used in food handling can be thermoformed from extruded sheet PLA, even yielding products with higher flex-crack resistance in living hinges (thin sections of plastic that bridge two parts) than those made of polystyrene. In addition, thin sheets of many PLA variants possess high gloss, excellent heat sealability, and clarity, allowing extruded thin sheets of PLA to replace cellophane and PET in transparent packaging [39]. Bags for yard and/or food wastes form another set of applications in which the physical properties as well as the biodegradable nature of PLA can be used to advantage; bags are tough and puncture resistant but are completely degraded within 4 to 6 weeks in a composting environment [39].

Extrusion coating of PLA for paper products is another set of applications with several benefits. For example, paper coated with PLA does not require pretreatment for ink adherence, whereas polyethylene (PE) paper coatings often do. PLA coatings also possess higher gloss, greater clarity, lower coefficients of friction, and greater stiffness than PE, allowing thinner paper to be used [18].

Some limitations do still exist, however, to the use of PLA in polystyrene-like applications. These include primarily the low-melt strength (extensibility without breaking of the molten state) and the relatively low temperature at which heat distortion begins to occur.

The drawbacks of PLA are overshadowed by its advantages in products with short lifecycles [38].

3.3. Synthesis

The two general synthetic routes for PLA include the condensation polymerization of lactic acid and the ring-opening polymerization of lactide. Condensation polymerization methods have produced high molecular weight PLA through chain extension and, more recently, through dehydration using azeotropic high-boiling (boiling point > 150°C) aprotic solvents and vacuum techniques. Polymerization utilizing a ring-opening mechanism is the preferred method of synthesis, however, and is the basis of the commercial Natureworks process.

Condensation polymerization has recently become a reliable method for production of high molecular-weight PLA. Attempts as recently as the mid-1990s to produce high molecular- weight PLA through simple dehydration without catalysis were unsuccessful, primarily due to side reactions [40]. To reduce the interfering side reactions, two methods are currently available. The first, developed by Ajioka et al. in 1995 [40], involves the dehydration of lactic acid using high-boiling aprotic solvents, such as diphenyl ether, that form azeotropes with water under vacuum conditions, as well as a catalyst. The second method involves using a dual catalyst system [41]. Other methods have been attempted with the goal of producing high molecular- weight PLA directly from lactic acid monomers, but to date have yielded only moderate molecular weight materials [42, 43].

Chain extension of oligomeric PLA is another approach, within the category of condensation polymerization, used to produce higher molecular weight PLA. Two basic methodologies have been developed. The first uses an additive to promote esterification of two PLA chains into one continuous chain [44, 45] while the other uses a linking agent to couple two or more chains together. Chain-linking agents such as diisocyanates [46, 47], thiirane [47], and diacidchlorides [44] are more economical than esterification promoting agents due to fewer purification steps and the important ability to run the reactions in the bulk. Problems associated with linking agents, however, include the persistence of unreacted chain-linking agents, residual metals, and the non-biodegradability of the linking agent, all of which diminish the environmental advantages of the resultant PLA.

Although the straight dehydration of lactic acid does not produce high molecular-weight polymers, the process is important in the production of lactide. The amount of lactide produced is influenced by temperature, pressure, and the types of catalysts added to the system [48]. Multiple purification methods are used industrially, including Cargill's multiple reflux controlled columns [11, 12], multi-step recrystallization reactors [49–56], and direct vapor-phase reaction of lactic acid [57, 58]. Formation of lactide occurs through a back-biting reaction involving two lactic acid monomers to form the six-member lactide ring. Three types of lactide can be formed: L-lactide or D-lactide with melting point = 97°C, or Meso-lactide with melting point = 52°C, depending on the composition of the starting material and extent of racemization (Figure 10). A fourth type, D,L-lactide, is formed when a racemic stereocomplex of D- and L-lactide are crystallized together; this form has a much higher melting point of 126–127°C [16].

Lactide, once formed, can be polymerized by three general mechanisms: anionic [33, 59-61], cationic [33, 62, 63], and coordination insertion (see Figure 11). Of these, coordination-insertion is the most prevalent and industrially most important.

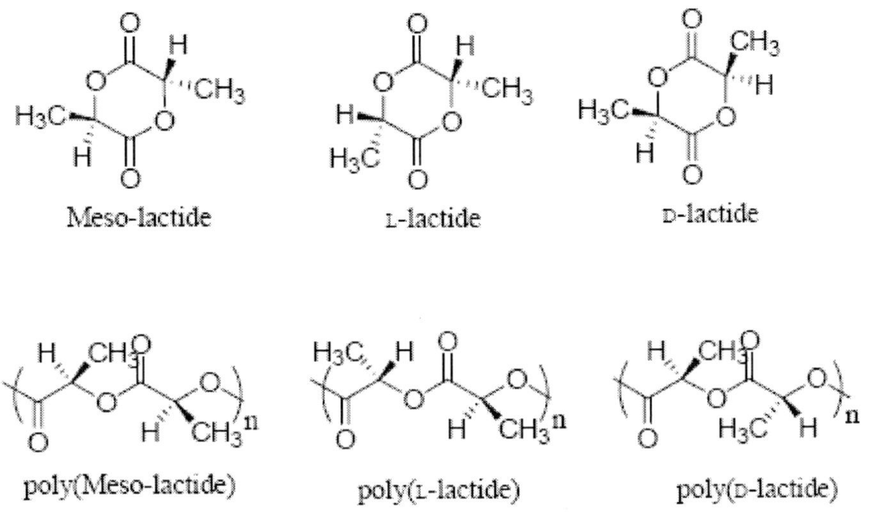

Figure 10. Types of lactide: Meso-lactide, L-lactide, and D-lactide.

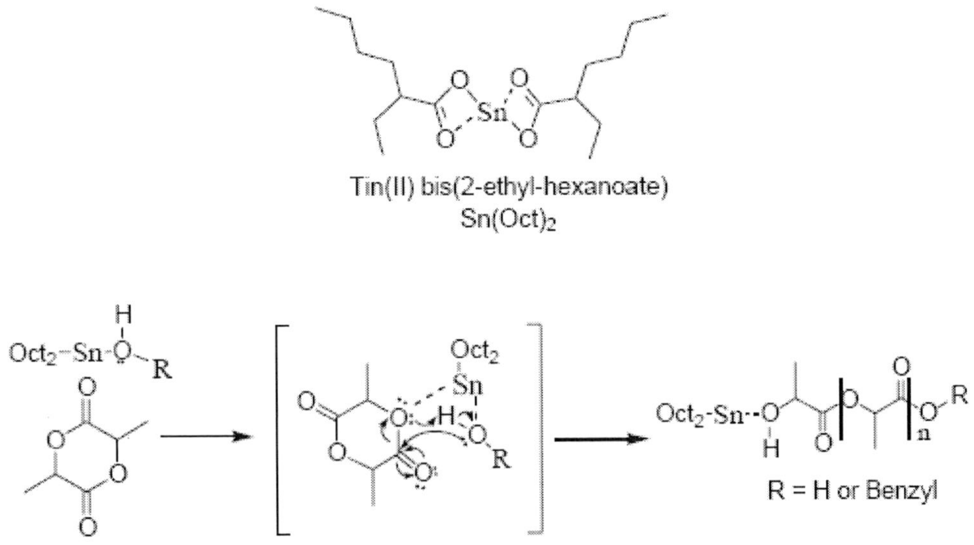

Figure 11. Coordination insertion polymerization of lactide with Sn(Oct)2 [73].

Polymerization reactions may be performed in bulk or in solution. The type of lactide, reaction temperature, and catalyst system determine the stereochemistry at each carbon-carbon bond along the backbone of PLA, in turn determining the properties of the resulting material. Impurities present in the lactide such as water, lactic acid, and lactyl-lactic acid decrease the molecular weight of the end polymer [64, 65].

The coordination insertion mechanism is the most common method used to produce high molecular weight PLA with high optical purity and crystallinity. Many catalytic or initiator systems have been developed for the coordination insertion polymerization of lactide. Oxides, halides, and alkoxides of metals possessing free p-, d-, or f- orbitals such as magnesium (Mg)

[66-68], zinc (Zn) [66, 69-73], aluminum (Al) (74-83), tin (Sn) [64, 65, 73, 80, 82, 84-94], and yittrium (Y) [95, 96] effectively catalyze the ring-opening polymerization of lactide. In general, the most useful catalysts possess highly covalent metal-ligand bonds [33].

The general catalytic mechanism involves coordination of the lactide carbonyl group to the catalyst metal, followed by cleavage of the ring acyl-oxygen bond and attachment of the growing, catalyst-attached polymer (Figure 11). The terminally bound catalyst then promotes the addition of successive monomers. Because each catalyst or initiator molecule facilitates the extension of a single polymer chain, molecular weights of PLA are controlled by varying the proportion of catalyst to lactide.

Among the coordination insertion catalysts, tin 2-ethylhexanoate (tin octoate, or Sn[Oct]2) is the most widely used and studied due to its ability to produce highly crystalline PLA in relatively short periods of time with high conversion and low racemization up to 180°C. It has also been approved by the U.S. Food and Drug Administration for food contact [16], making it ideal for many packaging applications Both kinetics and mechanisms of tin octoate polymerizations have been well characterized at both low [64] and high (≥1 80°C) temperatures [65]. Water, lactic acid, and lactyl lactate can also form other species with tin octoate, including tin oxide, tin hydroxides, tin lactate, all of which have been shown to catalyze the ring opening of lactic acid [65, 97, 98]. Lactide polymerized with tin octoate is best described by a second-order insertion mechanism [64] that is first order in monomer concentration [73].

Recently, single-site catalysts prepared by the addition of bulky side groups to metals have produced PLA with stereochemically-controlled structures [99]. Highly syndiotactic PLA has been formed from meso-lactide using bulky racemic aluminum catalysts [100]. Using the same catalyst, D,L-lactide was polymerized to form "stereoblocks" of PLA, possessing blocks of isotactic D- and L-PLA [101]. The stereoblock copolymer had a Tg slightly higher than isotactic L- or D-PLA, but lower than the stereocomplex formed between D- and L-PLA. Similar results were observed for the polymerization of D,L-lactide [102] using the same catalyst. Single-site catalyst work thus demonstrates a method to produce PLA with good control of the polymer backbone stereochemistry. This is a promising development, now opening the possibility of forming stereocomplexes between copolymer blocks that can withstand higher temperatures before heat distortion occurs at the interfaces between blocks.

3.4. Architectural Variations

Using techniques developed for polymerization of linear PLA, it has been possible to synthesize random copolymers, block copolymers, and branched polymers to modify the properties of the materials produced.

Random copolymers are produced by polymerization of two or more species together using a catalyst functional with both. Monomers that fit the criteria for ring-opening copolymerizations with lactide are glycolide [103-105], glycolide derivatives [106-109], lactones [30, 110-112], cyclic amide ethers [113], cyclic amide esters [114, 115], cyclic ether esters [31, 116, 117], cyclic phosphates [118] cyclic anhydrides (119) or cyclic carbonates [120-122]. Additional copolymers have been made through transesterification reactions of PLA with other polyesters, including PET [123]. By adjusting the proportions of each monomer, properties such as Tg, ease of biodegradation, and backbone flexibility can be altered [31, 117].

Block copolymers can also be produced. In one method, a ring-opening polymerization mechanism is employed using either coordination insertion catalysis or anionic catalysis [124]. Monomer addition sequence is important in this method, those that produce primary hydroxyl groups, such as lactones [113, 125-129], trimethylene carbonate (TMC) [122, 128], and 1,5- dioxepan-2-one (DXO) [130] must generally be polymerized first. Exceptions to the order of addition involve both glycolide [131] and D,L-3-methyl glycolide (MG) [109], which can be polymerized sequentially before or after PLA to form block copolymers because they possess the same type of propagating chain end as lactide. Ring-opening polymerization from polymeric initiators, a second method to produce block copolymers, employs polymeric initiators that possess hydroxyl groups on their chain ends. These groups initiate polymerization with the same catalysts used in polymerization of linear PLA. Effective polymeric initiators include poly(ethylene glycol) [124, 132-140], poly(ethylene glycol-co-propylene glycol) [141], poly(propylene glycol) [105], poly(tetrahydrofuran) [133, 142], poly(dimethyl siloxane) [133, 143-145], hydroxyl functionalized polyethylene [146], and hydroxy functionalized poly(isoprene) [147, 148]. Dendritic molecules with a single hydroxyl group have also been used to produce linear/dendritic di-block copolymers [149].

Several methods for synthesis of highly branched polymers through the ring opening of lactones have likewise been developed. Star formations are produced either by divergence, where polymer chains grow from a central core, or through convergence in which a termination reaction couples together individual chains [150-152], glycerol [133, 152-155]. Comb-shaped copolymers consisting of linear backbones with pendent linear chains may also be produced [156-162]. Lactides have also been polymerized into hyperbranched structures, in which many branch points lead to other branch points [155, 163-166]. Star-shaped copolymers are generally produced in a similar manner to star-shaped homopolymers, with hydroxyl groups on a multifunctional initiator used to initiate polymerization [167].

3.5. Fundamental Chain Properties

The characterization of PLA depends on the accurate knowledge of its fundamental properties. These include such parameters as Mark-Houwink constants that relate intrinsic viscosity to molecular weights, theta-condition front factors (K) used to calculate single chain properties, and characteristic ratios (C) that give an indication of the bonding structure of polymers. Unfortunately, each of these is reported inconsistently in the literature [5, 6, 150, 168- 174].

A recent study [175] has addressed these inconsistencies experimentally. PLA homopolymers and copolymers spanning wide ranges of molecular weight and stereoisomer proportion were prepared by ring-opening polymerizations of L-and D-lactides using tin octanoate as the catalyst. Samples were then characterized by means of dilute-solution viscometry in three different solvents; size-exclusion chromatography; static multiangle light scattering; variable-angle spectroscopic ellipsometry; and melt rheology. Data provided by these experiments include values of characteristic ratios as well as Schulz-Blaschke and MarkHouwink constants, all of which show consistently that polylactides are typical linear flexible polymers, in excellent agreement with recently published theoretical simulation results [176]

3.6. Processing Properties

Plastics are typically fabricated into useful articles in the molten state, thereby causing the melt flow, or rheological, properties of a polymer to be of great importance. As discussed above, PLAs have physical properties useful in fibers, packaging, and other applications traditionally dominated by petroleum-based resins. Although the general literature on polylactides is extensive, only a few articles [177-181] have considered rheological properties. Measurements of dynamic, steady, and transient shear viscosities have been presented and the extensional data on PLA showed a strong strain hardening behavior [181]. These studies did not, however, capture a systematic description of PLA rheology across a broad range of stereooptical composition, as materials studied to date have usually possessed high (>90 percent) *L-stereochemical center content*. However, a recent study [182] has provided a comprehensive evaluation of the linear viscoelastic properties of PLA across a wide range of molecular weights and stereochemical compositions. The rheology of blends of linear and branched-PLA architectures has also been comprehensively investigated [180, 183]. The results suggest that excellent control over rheological behavior of PLA is possible through blending chain architectures without compromising mechanical properties.

Mechanical properties of solution-spun [36, 37, 184, 185] and melt-spun [34, 36, 186-189] PLA fibers have been thoroughly investigated. It has been found that these properties are roughly equivalent to other polyesters meaning that PLA can replace textiles based on non-renewable resources. In addition, scanning electron microscopy (SEM) [34, 37, 184, 186] and wide-angle x-ray scattering (WAXS) [36, 190] have been useful in examining surface structure with respect to roughness and fracture surfaces. Understanding surface properties is important for dying and other textile finishing operations. Cicero et al. have provided a complete characterization of the hierarchical fiber morphology from linear PLAs [35, 191], determining thermal, mechanical, and morphological properties of the fibers and showing that properties can be widely manipulated through a combination of processing temperature and draw ratio (the amount of stretching the fiber undergoes). The same researchers have studied the effects of branching on fiber properties and morphology [192] and investigated the improvement of fiber properties specifically when thermally stabilized PLA is used [193]. These research studies provide several routes for optimizing the performance of PLA when used as a textile fiber in place of conventional, fossil resource based polyesters such at PET.

3.7. Permeation Properties

Because of the desirability of using PLA in packaging applications, understanding the permeation properties of PLA with respect to various gases and vapors, especially those of interest to the food industry, is extremely important. PLA is, unfortunately, a relatively poor barrier to water vapor and CO_2, with a water-vapor transmission rate significantly higher than those of PET, polypropylene (PP), or polyvinyl chloride (PVC), [194] resulting in some limited highly environmentally beneficial applications of PLA to packaging. For example, PLA's high permeability to water prevents it from being used to bottle water over long durations, despite the fact that increasing sales of bottled water have created considerable pressure on landfill space, particularly in California.

In response, Natureworks has developed several biaxially-oriented PLA films to improve PLA barrier properties, including two with trade names of PLA 4030-D and PLA 4040-D. Permeation of nitrogen, oxygen, CO_2, and CH4 in very thin (5 μm) amorphous films of various grades of PLA (L:D ratios from 95:5 to 98:2) cast from solution have also been examined [195]. Notably, changes in polymer chain branching and L:D ratios had no effect on the permeation properties of PLA with respect to small gases.

4. Research Priorities

4.1. Development of LCIA Tools

One of the greatest challenges facing the production of truly environmentally-benign plastic materials, PLA and otherwise, is the evaluation of net environmental impacts, beginning with feedstock production (agriculture or collection of biomass wastes), including processing steps (production of lactide and subsequent polymerization) and ending with the emissions resulting from biodegradation. Specific processes chosen at each stage, particularly concerning conventional vs. sustainable methods, are likely to have dramatic impacts on the net environmental profiles of individual materials, yet the tools with which to evaluate these differences are not yet fully refined [196]. These metrics are needed urgently, both to guide research and development of bioplastics and to advocate use of the truly environmentally beneficial materials. Consequently, a top priority in the development of environmentally benign plastics is the continuation of efforts to develop tools and standards within the context of LCIA that will make comparisons transparent and meaningful.

4.2. Improvement of Physical Properties

Presently, PLA and other biopolyesters suffer from two important deficiencies that limit their use. The first of these is their relatively low heat distortion temperatures, and the second is their relatively high permeabilities toward a number of substances, particularly water. As current, best-available LCIA analyses have indicated that PLA is indeed environmentally benign, continued research into biological, chemical, and physical transformations of PLA-based materials to improve these properties is warranted. In particular, nanocomposite technologies (Chapter III.D.) hold promise of improving both temperature distortion and permeation characteristics, as they have in conventional plastics, and should be investigated. Microcomposite technologies are related, already well-established approaches to achieve similar improvements in conventional plastics. In addition, plant microparticles derived from waste agricultural residues and simple grasses can be used directly as microparticles, providing both economic and environmental advantages [197, 198]. Alternatively, blending and trans-reacting PLA-based plastics with starch- or triglyceride-based materials (Chapter III.D.) may improve performance while maintaining biodegradability, with the result that these techniques also deserve further investigation [199-205]. Recently, copolymerization of cellulose acetate with PLA has demonstrated that the heat distortion temperature can be increased [206]. In this interesting case, both constituents of the plastic material come from renewable resources. This suggests that copolymerization of PLA, especially with other polymers based on renewable resources, can provide a viable route towards improved performance.

4.3. Exploration and Development of New Polyesters

A recent comprehensive study by the DOE [207] has identified 12 promising low molecular-weight materials that can be produced by fermentation in commercial quantities from plant sugars (succinic, fumaric, malic, 2,5-furandicarboxylic, 3-hydroxypropionic, aspartic, glutaric, glutamic, itonic, and levulinic acids, and the alcohols 3-hydroxybutyrolactone, glycerol, sorbitol, and xylitol). Combination of these acids and alcohols can produce polyesters by direct condensation [44, 208, 209]. In particular, reactive intermediates that can be produced by anaerobic fermentations are desirable, because anaerobic processes typically lose much less of the feedstock carbon to CO2 than do aerobic processes [210]. The success of the DuPont SoronaTM material, a polyester of such low molecular-weight precursors (1,4- benzenedicarboxylic acid-dimethyl ester with 1,3 - propanediol) shows that development of sustainable processes to take advantage of readily available, renewable substances to produce additional biodegradable plastics deserves high priority for its great potential to yield both homopolymeric and copolymeric materials with new ranges and combinations of desirable properties.

5. Commercialization

Polymers based on polylactic acid are the leading success stories in bioplastics, moving from the laboratory into the market within the last decade. PLA-based polymer synthesis is protected by numerous patents to several different entities [10, 47, 211-214] and Natureworks appears to have taken the lead in the United States with its opening in 2002 of the first commercial manufacturing facility for PLA in Blair, Nebraska. Natureworks continues to be the most internationally visible PLA producers, but Japan and Germany are showing interest in developing PLA commercially as well (http://www.friendlypackaging.org.uk /materialslist.htm).

Beyond issues of heat distortion and permeability described above, the greatest current obstacle to greater market penetration of PLA is cost. Increasingly, higher oil prices are enabling PLA to compete directly with petroleum-based plastics despite the fact that PLA is a new entity in the plastics marketplace, but cost reductions would nevertheless allow greater utilization. Accordingly, renewable materials such as starches, cellulose, and wood flours and fibers that could be used in PLA blends to decrease costs without significantly degrading performance are of great commercial interest. In this respect, bioplastics have a general advantage over biofuels– they are already clearly cost competitive with fossil-resource based materials.

REFERENCES

[1]   Carothers, W. H., G. L. Dorough, and F. J. Van Natta (1932). *Studies of polymerization and ring formation. X. The reversible polymerization of six-membered cyclic esters*, J of Am Chem Soc 54:761-772.

[2]   Lowe, C. E. (1954). *Preparation of High Molecular Weight Polyhydroxyacetic Ester*. U.S. Patent 2,668,162, February 2, 1954.

[3]   Goodman, M., and M. D'Alagni (1967). *Investigations of poly-S-lactic acid structure in solution*, Polymer Let 5:515-521.

[4] Schulz, R. C., and A. Guthmann (1967). *Optical rotatory dispersion and circular dichroism of poly-L-lactide.* A Reply to a Letter by M. Goodman and M. D'Alagni, Polymer Letters 5:1099-1102.

[5] Tonelli, A. E., P. J. Flory (1969). *The configuration statistics of random poly (lactic acid) chains. I. Experimental results,* Macromolecules 2(3) :225-227.

[6] Tonelli, A. E., P. J. Flory, and D. A. Brant (1969). *The configurational statistics of random poly (lactic acid) chains.* II. Theory, Macromolecules 2:227.

[7] Schulz, V. R. C., and J. Schwaab (1965). *Rotationsdispersion bei Monomerem und Polymerem L-(-)-Lactid,* Die Makromolekulare Chemie 87:90-102.

[8] Kulkarni, R. K., K. C. Pani, C. Neuman, and F. Leonard (1966). *Polylactic acid for surgical implants,* Arch of Surg 93:839-843.

[9] Middleton, J. C., and A. J. Tipton (1998). *Synthetic biodegradable polymers as medical devices.* Med Plastics and Biomaterials Mag 30.

[10] Gruber, P. R., E. S. Hall, J. J. Kolstad, M. L. Iwan, R. D. Benson, and R. L. Borchardt (1993). *Continuous Process for Manufacture of Lactide Polymers with Controlled Optical Purity.* U.S. Patent 5,247,058, September 21.

[11] Gruber, P. R., E. S. Hall, J. J. Kolstad, M. L. Iwen, R. D. Benson, and R. L. Borchardt (1993). *Continuous Process for the Manufacture of a Purified Lactide from Esters of Lactic Acid.* U.S. Patent 5,247,059, September 21.

[12] Gruber, P. R., E. S. Hall, J. J. Kolstad, M. L. Iwen, R. D. Benson, and R. L. Borchardt (1993). *Continuous Process for Manufacture of a Purified Lactide.* U.S. Patent 5,274,073, December 28.

[13] Gruber, P. R., E. S. Hall, J.J Kolstad, M. L. Iwen, R. D. Benson, R. L. Borchardt (1994). *Continuous Process for Manufacture of Lactide Polymers with Purification by Distillation.* U.S. Patent 5,357,035, October 19.

[14] Vert, M., G. Schwarch, and J. Coudane (1995). *Present and Future of PLA Polymers,* J.M.S.-Pure Appl Chem A32(4):787-796.

[15] Dorgan, J. R. (1999). Poly (lactic acid) Properties and Prospects of an Environmentally Benign Plastic, *3rd Annual Green Chemistry and Engineering Conference Proceedings,* Washington, D.C., pp 145-149.

[16] Hartman, M. H. (1998). *High Molecular Weight Polylactic Acid Polymers, in Biopolymers from Renewable Resources* (D. H. Kaplan, Ed.). Springer-Verlag, Berlin, pp 367-411.

[17] Naitove, M. H. (1995). *Push is on to commercialize biodegradable lactide polymers.* Plas Tech, March 15-17.

[18] Riggle, D. (1997). Composting yard bags: in Minnesota, Biocycle, Journal of Composting and Recycling 38(2): 65.

[19] Zhang, X., U. P. Wyss, D. Pichora, and M. F. A. Goosen (1994). *An investigation of poly (lactic acid) degradation,* J of Bioactive and Compatible Polymers 9:81-100.

[20] Vink, E. T. H., K. R. Ra´ bago, D. A. Glassner, and P. R. Gruber (2003). *Applications of life cycle assessment to Nature Works polylactide* (PLA) production, Polymer Degradation and Stab 80:403-419.

[21] Lu, L., and A. G. Mikos (1999). Poly(lactic acid), in *Polymer Data Handbook* (J. E. Mark, Ed.). Oxford University Press, New York, pp 627-63 3.

[22] Engelberg, I., and J. Kohn (1991). *Physico-mechanical properties of degradable polymers used in medical applications: a comparative study*, Biomaterials 12(5):292-304.

[23] Ikada, Y., K. Jamshidi, H. Tsuji, and S. H. Hyon (1987). *Stereocomplex formation between enantiomeric poly(lactides)*, Macromolecules 20:904-906.

[24] Vainionpää, S., P. Rokkanen, P. Törmälä (1989). Surgical applications of biodegradable polymers in human tissues, Progress in Polymer Sci 14:679-708.

[25] Eitenmüller, J., A. David, A. Pommer, and G. Muhr (1996). Operative hehandlung von sprunggelenksfrakturen mit biodegradablem schrauben und platten aus poly-L-lactide, Der Chirurg 67:413-418.

[26] Viljanen, J. T., H. K. Pihlajamäki, P. O. Törmälä, and P. U. Rokkanen (1997). Comparison of the tissue response to absorbable self-reinforced polylactide screws and metallic screws in the fixation of cancellous bone osteotomies: *An experimental study on the rabbit distal remur*, Journal of Orthopaedic Research 15:398-407.

[27] Richardson, J. B., R. Jones, A. Hunt, and S. J. Brat (2000). *The BiolokTM Screw A New Generation in Resorbables*, in Orthopeadic Product News: Online Edition, Ed. http://www.opnews.com/ole/archives/articles/mar.may00/items.html.

[28] Pétas, A., M. Talja, T. L. J. Tammela, K. Taari, T. Välimaa, and P. Törmälä (1997). The biodegradable self-reinforced poly-DL-lactic acid spiral stent compared with a suprapubic catheter in the treatment of post-operative urinary retension after visual Laser ablation of the prostate, British Journal of Urology 80:43 9-443.

[29] Vert, M. (1986). *Biomedical polymers from chiral lactides and functional lactones: properties and applications*, Die Makromolekulare Chemie, Macromolecular Symposia 6: 109-122.

[30] Ge, H., Y. Hu, S. Yang, X. Jiang, and C. Yang (2000). *Preparation, characterization, and drug release behaviors of drug-loaded e-Caprolactone/L-Lactide copolymer nanoparticles*, Journal of Applied Polymer Science 75(7):874-882.

[31] Edlund, U., and A. C. Albertsson (1999). *Novel drug delivery microspheres from Poly(1,5-dioxepan-2-one-co-L-lactide)*, Journal of Polymer Science, Part A: Polymer Chemistry 37:1877-1884.

[32] Griffith, L. G. (2000). *Polymeric biomaterials*, Acta Materialia 48:263-277.

[33] Kricheldorf, H. R., I. Kreiser-Saunders, C. Jürgens, and D. Wolter (1996*). Polylactides - synthesis, characterization and medical application*, Macromolecular Symposium 103 :85-102.

[34] Yuan, X., A. F. T. Mak, K. W. Kwok, B. K. O. Yung, and K. Yao (2001). *Characterization of poly(L-lactic acid) fibers produced by melt spinning*, Journal of Applied Polymer Science 81(1) :251-260.

[35] Cicero, J. A., and J. R. Dorgan (2001). *Physical properties and fiber morphology of poly (lactic acid) obtained from continuous two-step melt spinning*, J Polymers and Environ (January): 1-15.

[36] Eling, B. S. Gogolewski, and A. J. Pennings (1982). *Biodegradable Materials of Poly(Llactic acid): 1. Melt Spun and Solution-Spun Fibers*, Polymer 23:1587-1593.

[37] Fambri, L., A. Pegoretti, and C. Migliaarsesi (1994). *Biodegradable fibres. Part I. Poly-L -lactic acid fibres produced by solution spinning*, Journal of Material Science, Materials in Medi 5(9/10):679-683.

[38] Rudie, R. (2000). *Cargill Dow Sows Seeds of Future Fibers*; Will Build $300 Million PLA Polymer Plant, International Fiber Journal, February.

[39] Auras, R. A., B. Harte, S. Selke, and R. J. Hernandez (2002). In *Mechanical, Physical, and Barrier Properties of Poly (lactic acid) Films*, Worldpak-IAPRI, East Lansing, MI, 48824-1224.

[40] Ajioka, M., K. Enomoto, K. Suzuki, and A. Yamaguchi (1995). *Basic properties of polylactic acid produced by the direct condensation polymerization of lactic acid*, Bulletin of the Chemical Society of Japan 68:2125-2131.

[41] Moon, S. I., C. W. Lee, M. Miyamoto, and Y. Kimura (2000). *Melt polycondensation of L-lactic acid with Sn(II) catalysts activated* by various proton acids: A direct manufactoring route to high molecular weight poly(L-lactic acid), Journal of Polymer Science, Part A: Polymer Chemistry 38:1673-1679.

[42] Otera, J., K. Kawada, and T. Yano (1996). *Direct condensation polymerization of l-lactic acid catalyzed by distannoxane,* Chemistry Letters 3:225-226.

[43] Hiltunen, K., J. V. Seppal, and M. Harkonen (1997). *Effect of catalyst and polymerization conditions on the preparation of low molecular weight lactic acid polymers*, Macromolecules 30:373-379.

[44] Buchholz, B. (1994). *Process for Preparing Polyesters Based on Hydroxycarboxylic Acids.* U.S. Patent 5,302,694, April 12.

[45] Aharoni, S. M., and D. Masilamani (1986). *Process for Preparing Extended Chain Polyesters and Block or Graft Copolyesters.* U.S. Patent 4,568,720, February 4.

[46] Woo, S. I., B. O. Kim, H. S. Jun, and H. N. Chang (1995). *Polymerization of aqueous lactic acid to prepare high molecular weight poly (lactic acid) by chain-extending with hexamethylene diisocyanate*, Polymer Bulletin 35:415-421.

[47] Bonsignore, P. V. (1995). *Production of High Molecular Weight Polylactic Acid.* U.S. Patent 5,470,944, November 28.

[48] Bellis, H. E. (1988). *Process for preparing highly pure cyclic esters.* U.S. Patent 4,727,163, February 23.

[49] Bhatia, K. K. (1991). *Continuous process for rapid conversion of oligomers to cyclic esters.* U.S. Patent 5,023,349, June 11.

[50] Bhatia, K. K. (1991). *Process for the purification of cyclic esters.* U.S. Patent 5,023,350, June 11.

[51] Bhatia, K. K. (1991). *One-step continuous process for preparing cyclic esters.* U.S. Patent 5,043,458, August 27

[52] Bhatia, K. K. (1992). *Process for rapid conversion of oligomers to cyclic esters.* U.S. Patent 5,091,544, February 25.

[53] Bhatia, K. K. (1992). *Solvent scrubbing recovery of lactide and other dimeric cyclic esters.* U.S. Patent 5,266,706, November 30.

[54] Bhatia, K. K., N. E. Drysdale, J. R. Kosak (1992). *Solvent scrubbing recovery of lactide and other dimeric cyclic esters.* U.S. Patent 5,117,008, May 26.

[55] Bhatia, K. K., J. V. Tarbell (1993). *Co-vaporization process for preparing lactide.* U.S. Patent 5,196,551, March 23.

[56] Fridman, I. D., and J. Kwok (1993). *Lactide Melt Recrystallization.* U.S. Patent 5,264,592.

[57] Benecke, H. P., A. Cheung, G. E. Cremeans, M. E. D. Hillman, E. S. Lipinsky, R. A. Markle, and R. G. Sinclair (1994). *Method to produce cyclic esters*. U.S. Patent 5,319,107, June 7.

[58] Sinclair, R. G., R. A. Markle, and R. K. Smith (1993). *Lactide production from dehydration of aqueous lactic acid feed.* U.S. Patent 5,274,127, December 28.

[59] Kleine, V. J., and H. H. Kleine (1959). Über hochmolekulare, insbesondere optisch aktive polyester der milchsäure, ein beitrag zur stereochemie makromolekularer verbindungen, Die Makromolekulare Chemie 30:23-3 8.

[60] Kricheldorf, H. R., and I. Kreiser-Saunders (1990). Anionic polymerization of L-lactide in solution, Die Makromolekulare Chemie 191:1057-1066.

[61] Kricheldorf, H. R., and C. Boettcher (1993). *Polylactones 26: Lithium alkoxide-initiated polymerizations of L-lactide*, Die Makromolekulare Chemie 194(6): 1665-1669.

[62] Kricheldorf, H. R., and Dunsing (1986). Mechanism of the cationic polymeriation of L,Ldilactide, Die Makromolekulare Chemie 187:1611-1625.

[63] Dittrich, W., and R. C. Schulz (1971). *Kinetics and mechanism of the ring-opening polymerization of L(–)-lactide*, Angew. Makromol Chem 15:109-126.

[64] Kricheldorf, H. R., I. Kreiser-Saunders, and C. Boettcher (1995). *Polylactones 31. Sn (II)octoate-initiated polymerization of l-lactide: A mechanistic study*, Polymer 36(6):1253-1259.

[65] Kricheldorf, H. R., I. Kreiser-Saunders, and A. Stricker (2000). Polylactones 48. SnOct2-initiated polymerization of lactide: A mechanistic study, Macromolecules 33:702-709.

[66] Kricheldorf, H. R., I. Kreiser-Saunders, and D. O. Damrau (2000). *Resorbable initiators for polymerization of lactones,* Macromolecular Symposium 159(1) :247-258.

[67] Kricheldorf, H. R., M. Lossin (1997). *Polylactones 34: Polymerizations of meso- and rac-D,L,-lactide by means of grignard reagents,* Journal of Macromolecular Science, Pure and Applied Chemistry 34(1):179-189.

[68] Kricheldorf, H. R., S.-R. Lee (1995). *Polylactones 32: High-molecular-weight polylactides by ring-opening polymerization with bibutylmagnesium or butylmagnesium chloride*, Polymer 36(15):299-307

[69] Kricheldorf, H. R., and D. O. Damrau (1997). *Polylactones 37: Polymerizations of L-lactide initiated with Zn(II) L-lactate and other resorbable Zn salts*, Macromolecular Chemistry and Physics 198:1753-1766.

[70] Kricheldorf, H. R., and D. O. Damrau (1998). *Polylactones 43: Polymerization of L-lactide catalyzed by zinc amino acid salts, Macromolecular Chemistry and Physics* 199(8):1747-1752.

[71] Kricheldorf, H. R., and D. O. Damrau (1998). *Polylactones 42: Zn L-lactate-catalyzed polymerizations of 1,4-dioxan-2-one,* Macromolecular Chemistry and Physics 199(6):1089-1097.

[72] Kricheldorf, H. R., and C. Boettcher (1993). *Polylactones XXV. Polymerizations of racemic-and meso-D,L-lactide with Zn, Pb, Sb, and Bi salts-stereochemical aspects, Journal of Macromolecular Science,* Pure and Applied Chemistry 30(6/7):441-448.

[73] Nijenhuis, A. J., D. W. Grijpma, and A. J. Pennings (1992). *Lewis acid catalyzed polymerization of L-lactide. kinetics and mechanism of the bulk polymerization,* Macromolecules 25:6419-6424.

[74] Huang, C.-H., F. C. Wang, B. T. Ko, T. L Yu, and C. C. Lin (2001). *Ring opening polymerization of e-caprolactone and L-lactide using aluminum thiolates as initiators*, Macromolecules 34:356-361.

[75] Cameron, P. A., D. Jhurry, V. C. Gibson, A. J. P. White, D. J. Williams, and S. Williams (1999). *Controlled polymerization of lactides at ambient temperature using [5-Clsalen]AlOMe*, Macromolecular Rapid Communications 20(12): 616-618.

[76] Emig, N., H. Nguyen, H. Krautscheid, R. Réau, J.-B. Cazaux, and G. Bertrand (1998).

[77] Neutral and cationic tetracoordinated aluminum complexes featuring tridentate nitrogen donors: Synthesis, structure, and catalytic activity for the ring-opening polymerization of propylene oxide and (D,L)-lactide, Organometallics 17(16): 3599-3608.

[78] Dubois, P., C. Jacobs, R. Jérôme, and P. Teyssié (1991). *Macromolecular engineering of polylactones and polylactides. 4. Mechanism and kinetics of lactide homopolymerization by aluminum isopropoxide*, Macromolecules 24:2266-2270.

[79] Eguiburu, J. L., and M. J. Fernandez Berridi (1995). *Functionalization of poly(L-lactide) macromonomers by ring-opening polymerization of L-lactide initiated with hydroxyethyl methacrylate-aluminum alkoxides,* Polymer 36(1): 173-179.

[80] Kricheldorf, H. R., M. Berl, and N. Scharnagl (1988). *Polylactones 9: Polymerization mechanism of metal alkoxide initiated polymerizations of lactide and various lactones*, Macromolecules 21:286-293.

[81] Kohn, F. E., J. G. Van Ommen, and J. Feijen (1983) T*he mechanism of the ring-opening polymerization of lactide and glycolide,* European Polymer Journal 19(12): 1081-1088.

[82] Eguiburu, J. L., M. J. Fernandex-Berridi, and J. San Roamn (1999). *Ring-opening polymerization of L-lactide initiated by (2-Methacryloxy)ethyloxy-Alum inumtTrialkoxides. 1. Kinetics*, Macromolecules 32(25): 8252-8258.

[83] Kricheldorf, H. R., and A. Serra (1985) *Polylactones 6. Influence of various metal salts on the optical purity of poly(L-lactide),* Polymer Bulletin 14:497-502.

[84] Montaudo, G., M. S. Montaudo, C. Puglisi, F. Samperi, N. Spassky, A. LeBorgne, and M. Wisniewski (1996). Evidence of ester-exchange reactions and cyclic oligomer formation in the ring-opening polymerization of lactide with aluminum complex initiators, Macromolecules 29:6461-6465.

[85] Kricheldorf, H. R., and S. Eggerstedt (1999). *Macrocycles 5: Ring expansion of macrocyclic tin alkoxides with cyclic anhydrides,* Macromolecular Chemistry and Physics 200(3):587-593.

[86] Kohn, F. E., J. W. A. V. D. Berg, G. V. D. Ridder, and J. Feijen (1984). *The ring-opening polymerization of D,L-lactide in melt initiated with tetraphenyltin,* Journal of Applied Polymer Science 29:4265-4277.

[87] Kricheldorf, H. R., S.-R. Lee, and S. Bush (1996). Polylactones 36: Macrocyclic polymerization of lactides with cyclic Bu2Sn initiators derived from 1,2-ethanediol, 2-mercaptoethanol, and 1,2-dimercaptoethane, Macromolecules 29(5):1375-1381.

[88] Möller, M., R. Kånge, and J. L. Hedrick (2000). *Sn(OTf)2 and Sc(OTf)3: Efficient and versatile catalysts for the controlled polymerization of lactones,* Journal of Polymer Science, Part A: Polymer Chemistry 38:2067-2074.

[89] Kowalski, A., J. Libiszowski, A. Duda, and S. Penczek (2000). *Polymerization of L,L dilactideInitiated by tin(II) butoxide,* Macromolecules 33(6): 1964-1971.

[90] Schindler, A., and K. D. Gaetano (1988). *Poly(lactate) III. Stereoselective polymerization of meso-dilactide*, Journal of Polymer Science, Part C: Polymer Letters 26:47-48.

[91] Kricheldorf, H. R., and M. Sumbél (1989). *Polylactones-18. Polymerization of L,L-lactide with Sn(II) and Sn(IV) halogenides*, European Polymer Journal 25(6):585-591.

[92] Kricheldorf, H. R., C. Boettcher, and K. U. Tönnes (1992). *Polylactones 23: Polymerization of racemic and meso D,L-lactide with various organotin catalysts-stereochemical aspects*, Polymer 33(1 3):28 17-2824.

[93] Dahlmann, J., and G. Rafler (1993). *Biodegradable Polymers*. 7th comm. On the mechanism of ring-opening polymerization of cyclic esters of aliphatic hydroxycarboxylic acids by means of different tin compounds, Acta Polymerica 44:103-107.

[94] Leenslag, J. W., and A. J. Pennings (1987). *Synthesis of high-molecular-weight poly (L-lactide) initiated with tin 2-ethylhexanoate*, Die Makromolekulare Chemie 188:1809-1814.

[95] Witzke, D. R., R. Narayan, and J. J. Kolstad (1997). *Reversible kinetics and thermodynamics of the homopolymerization of L-lactide with 2-Ethylhexanoic acid Tin(II) salt*, Macromolecules 30:7075-7085.

[96] Aubrecht, K. B., K. Chang, M. A. Hillmyer, and W. Tolman (2001). *Lactide polymerization activity of alkoxide, phenoxide, and amine derivative of Yttrium(III) arylamidinates*, Journal of Polymer Science, Part A: Polymer Chemistry 39:284-293.

[97] Chamberlain, B. M., B. A. Jazdzewski, M. Pink, M. A. Hillmyer, and W. B. Tolman (2000). *Controlled polymerization of D,L-lactide and e-caprolactone by structurally well-defined alkoxo-bridged di- and triyttrium (III) complexes*, Macromolecules 33:3970- 3977.

[98] Schwach, G., J. Coudane, R. Engel, and M. Vert (1997). *More about the polymerization of lactides in the presence of stannous octoate*, Journal of Polymer Science, Part A: Polymer Chemistry 35:3431-3440.

[99] Storey, R. F., and A. E. Taylor (1998). *Effect of stannous octoate on the composition, molecular weight, and molecular weight distribution of ethylene glycol-Initiated poly(e-caprolactone)*, Journal of Macromolecular Science, Pure and Applied Chemistry 35(5):723-750.

[100] Cheng, M., A. B. Attygalle, and G. W. Coates (1999). *Single-site catalyst for ring-opening polymerization: synthesis of heterotactic poly (lactic acid) from rac-lactide*, Journal of the American Chemical Society 121(49):1 1583-11584.

[101] Ovitt, T. M., and G. W. Coates (1999). *Stereoselective ring-opening polymerization of meso-lactide: synthesis of syndiotactic poly(lactic acid)*, Journal of the American Chemical Society 121:4072-4073.

[102] Ovitt, T. M., and G. W. Coates (2000). *Stereoselective ring-opening polymerization of rac-lactide with a single-site, racemic aluminum alkoxide catalyst: synthesis of stereoblock poly (lactic acid)*, Journal of Polymer Science, Part A: Polymer Chemistry 38:4686-4692.

[103] Radano, C. P., G. L. Baker, and M. R. I. Smith (2000). *Stereoselective polymerization of a racemic monomer with a racemic catalyst: direct preparation of the polylactic acid stereocomplex from racemic lactide*, Journal of the American Chemical Society 122: 1552-1553.

[104] Joziasse, C. A. P., H. Grablowitz, and A. J. Pennings (2000). *Star-shaped poly[trimethylene carbonate) -co-(e-capro-lactone)] and its block copolymers with lactide/glycolide: synthesis, characterization and properties*, Macromolecular Chemistry and Physics 201(1):107-112.

[105] Rafler, G., and J. Dahlmann (1990). *Biolgisch abbaubare polymere 2. Mitt.: Zur homo- und copolymerisation von D,L-dilactide*, Acta Polymerica 41(12):611-617.

[106] Farnia, S. M. F., M.-R. J., and M. N. Sarbolouki (1999). *Synthesis and characterization of novel biodegradable triblock copolymers from L-lactide, glycolide, and PPG*, Journal of Applied Polymer Science 73(5): 633-637.

[107] Hile, D. D., and M. V. Pishko (1999). *Ring-opening precipitation polymerization of poly(D,L-lactide-co-glycolide) in supercritical carbon dioxide*, Macromolecular Rapid Communications 20(10):51 1-5 14.

[108] Yin, M., and G. L. Baker (1999). *Preparation and characterization of substituted polylactides*, Macromolecules 32(23) :7711-7718.

[109] Dong, C.-M., K.-Y. Qiu, Z.-W. Gu, and X. D. Feng (2000). *Synthesis of poly(D,L-lactic acid-alt-glycolic acid) rrom D,L-3-Methylglycolide*, Journal of Polymer Science, Part A: Polymer Chemistry 38:4179-4184.

[110] Dong, C.-M., K.-Y. Qiu, Z.-W. Gu, and X. D. Feng (2001). *Living polymerization of D,L-3-Methylglycolide initiated with bimetallic (Al/Zn) u-Oxo alkoxide and copolymers thereof,* Journal of Polymer Science, Part A: Polymer Chemistry 39:357-3 67.

[111] Nakayama, A., N. Kawasaki, Y. Maeda, I. Aravanitoyannis, S. Aiba, and N. Yamamoto (1997). *Study of biodegradability of poly(d-valerolactone-co-lactide)s*, Journal of Applied Polymer Science 66(4) :741-748.

[112] Kasperczyk, J., and M. Bero (1993). *Coordination polymerization of lactides, the role of transesterification in the copolymerization of l,l-lactide*, Makromol Chem 194:913-925.

[113] Hiljanen-Vainio, M., T. Karjalainen, and J. Seppälä (1996). *Biodegradable lactone copolymers. I. Characterization and mechanical behavior of e-caprolactone and lactide copolymers*, Journal of Applied Polymer Science 59(8): 1281-1288.

[114] Deng, X., J. Yao, Y. Minglong, X. Li, and C. Xiong (2000). *Polymerization of lactides and lactones, 12. Synthesis of poly[(glycolic acid)-alt-(L-glutamic acid)] and poly {(lactic acid)-co-[glycolic acid)-alt-(L-glutamic acid)]}*, Macromolecular Chemistry and Physics 201(17):2371-2376.

[115] Ouchi, T., T. Nozaki, A. Ishikawa, I. Fujimoto, and Y. Ohya (1997). *Synthesis and enzymatic hydrolysis of lactic acid-depsipeptide copolymers with functionalized pendant groups*, Journal of Polymer Science, Part A: Polymer Chemistry 35(2):377-383.

[116] Ouchi, T., H. Seike, T. Nozaki, and Y. Ohya (1998*). Synthesis and characteristics of polydepsipeptide with pendant thiol groups*, Journal of Polymer Science, Part A: Polymer Chemistry 36:1283-1290.

[117] Albertsson, A.-C., U. Edlund, and K. Stridsberg (2000). *Controlled ring-opening polymerization of lactones and lactides*, Macromolecular Symposium 157(1):39-46.

[118] Stridsberg, K., M. Gruvegard, and A.-C. Albertsson (1998). *Ring-opening polymerization of degradable polyesters*, Macromolecular Symposium 130:367-378.

[119] Wen, J., and R. X. Zhuo. (1998). *Preparation and characterization of poly(D,L-lactide co-ethylene methyl phosphate)*, Polymer International 47(4) :503-509.

[120] Zhu, K. J., Y. Lei (1997). *Preparation, Characterization and Biodegradation Characteristics of Poly(adipic anhydride-co-D,L-lactide),* Polymer International 43(3):210-216.

[121] Keul, H., and H. Höcker (2000*). Expected and unexpected reactions in ring-opening (co)polymerization,* Macromolecular Rapid Communications 21(13): 869-883.

[122] Storey, R. F., B. D. Mullen, and K. M. Melchert (2001). *Synthesis of novel hydrophilic copolymers containing L,L-lactide and 5-methyl-5-benzyloxycarbonyl-1, 3-dioxane-2-one,* Polymer Preprints 42(1):553-554.

[123] Kricheldorf, H. R., and A. Stricker (1999). *Polylactones 47: A-B-A Triblock copolyesters and random copolyesters of trimethylene carbonate and various lactones via macrocyclic polymerization,* Macromolecular Chemistry and Physics 200(7): 1726-1733.

[124] Haderlein, G., C. Schmidt, J. H. Wendorff, and A. Greiner (1997*). Synthesis of hydrolytically degradable aromatic polyesters with lactide moieties,* Polymers for Advanced Technologies 8(9): 568-573.

[125] Deng, X., Z. Zhu, C. Xiong, and L. Zhang (1997). *Synthesis and Characterization of Biodegradable Block Copolymers of e-Caprolactone and D,L-Lactide Initiated by Potassium (ethylene glycol) ate,* Journal of Polymer Science, Part A: Polymer Chemistry 35:703-708.

[126] Qing, C., J. Bei, and S. Wang (2000). *Synthesis and characterization of polycaprolactone (B)-poly(lactide-co-glycolide) (A) ABA block copolymer,* Polymers for Advanced Technologies 11(4): 159-166.

[127] Song, C. X., and X. D. Feng (1984). *Synthesis of ABA triblock copolymers of e-caprolactone and DL-lactide,* Macromolecules 17:2764-2767.

[128] In't Veld, P. J. A., E. M. Velner, P. Van De Witte, J. Hamhuis, P. J. Dijkstra, and J. Feijen (1997). *Melt block copolymerization of e-caprolactone and L-lactide,* Journal of Polymer Science, Part A: Polymer Chemistry 35:219-226.

[129] Grijpma, D. W., R. D. A. Van Hofslot, H. Supèr, A. J. Nijenhuis, and A. J. Pennings (1994). *Rubber toughening of poly(lactide) by blending and block copolymerization,* Polymer Engineering and Science 34(22): 1674-1684.

[130] Jacobs, C., P. Dubois, and R. Jerome (1991). Macromolecular engineering of polylactones and polylactides. 5. synthesis and characterization of diblock copolymers based on poly-e-caprolactone and poly(L,L or D,L)lactide by aluminum alkoxides, Macromolecules 24(11): 3027-3034.

[131] Stridsberg, K., and A. C. Albertsson (2000). *Controlled ring-opening polymerization of L-lactide and 1,5-Dioxepan-2-one forming a triblock copolymer,* Journal of Polymer Science, Part A: Polymer Chemistry 38:1774-1784.

[132] Barakat, I., P. Dubois, C. Grandfils, and R. Jérôme (2001). *Poly(e-caprolactone-b-glycolide) and poly (D,L-lactide-b-glycolide) diblock copolyesters: Controlled synthesis, characterization, and colloidal dispersions,* Journal of Polymer Science, Part A: Polymer Chemistry 39:294-306.

[133] Kricheldorf, H. R., and C. Boettcher (1993). Polylactones 27. Anionic polymerization of L-lactide. Variation of endgroups and synthesis of block copolymers with poly(ethylene oxide), Die Makromolekulare Chemie, Macromolecular Symposia 73:47-64.

[134] Spinu, M., C. Jackson, M. Y. Keating, and K. H. Gardner (1996). *Material design in poly (lactic acid) systems: block copolymers, star homo- and copolymers, and stereocomplexes,* Journal of Macromolecular Science, Pure and Applied Chemistry 33(10):1497-1530.

[135] Cho, K. Y., C.-H. Kim, J.-W. Lee, and J. K. Park (1999). *Synthesis and characterization of poly(ethylene glycol) grafted poly (L-lactide),* Macromolecular Rapid Communications 20(1 1):598-601.

[136] Fujiwara, T., Y. Kimura, and I. Teraoka (2000). Tailoring of block copolymers based on the stoichiometric control of the end-functionality of telechilic oligomers and the utilization of large-scale fractionation by phase fluctuation chromatography: *A synthetic strategy for the preparation of end-functionalized poly(L-lactide)-block-poly(oxyethylene),* Journal of Polymer Science, Part A: Polymer Chemistry 3 8:2405-2414.

[137] Salem, A. K., S. M. Cannizzaro, M. C. Davies, S. J. B. Tendler, C. J. Roberts, P. M. Williams, and K. M. Shakesheff (2001). *Synthesis and characterization of a degradable poly (lactic acid) -poly (ethylene glycol) copolymer with biotinylated end groups,* Biomacromolecules 2:575-580.

[138] Kricheldorf, H. R., and J. Meier-Haack (1993). *Polylactones, 22 ABA triblock copolymers of L-lactide and poly (ethylene glycol),* Die Makromolekulare Chemie 194: 715-725.

[139] Zhu, K. J., L. Xiangzhou, and Y. Shilin (1990). *Preparation, characterization, and properties of polylactide (PLA) -Poly (ethylene glycol) (PEG) copolymers: A potential drug carrier,* Journal of Applied Polymer Science 39:1-9.

[140] Du, Y. J., P. J. Lemstra, A. J. LNijenhuis, H. A. M. v. Aert, and C. Bastiaansen (1995). *ABA type copolymers of lactide with poly (ethylene glycol). Kinetic, mechanistic, and model Studies,* Macromolecules 28(7):2124-2132.

[141] Yuan, M., Y. Wang, X. Li, C. Xiong, and X. Deng (2000). *Polymerization of lactides and lactones. 10. Sythesis, characterization, and application of amino-terminated poly (ethylene glycol)-poly(e-caprolactone) block copolymer,* Macromolecules 33:1613 -1617.

[142] Yamaoka, T., Y. Takahashi, T. Ohta, M. Miyamoto, A. Murakami, and Y. Kimura (1999). Synthesis and properties of multiblock copolymers consisting of poly(l-lactic acid) and poly (oxypropylene-co-oxyethylene) prepared by direct polycondensation, Journal of Polymer Scince, Part A: Polymer Chemistry 37:1513-1521.

[143] Kricheldorf, H. R., and D. Langanke (1999). *Macrocycles 8: Multiblock copoly (ether-esters) of poly(THF) and e-caprolactone via macrocyclic polymerization,* Macromolecular Chemistry and Physics 200(5):1 183-1190.

[144] Zhang, S., Z. Hou, and K. E. Gonsalves (1996). *Synthesis and characterization of L-lactide/a, w-hydroxypropyl terminated PDMS in the presence of Sn (II) 2-ethylhexanoate,* Polymer Preprints 38(1):853-854.

[145] Zhang, S., Z. Hou, and K. E. Gonsalves (1996). *Copolymer synthesis of poly(L-lactide-bDMS-L-lactide) via the ring opening polymerization of L-lactide in the presence of a,whydroxypropyl-terminated PDMS,* Journal of Polymer Science, Part A: Polymer Chemistry 34:2737-2742.

[146] Riffle, J. S., W. P., Jr. Steckle, K. A. White, and R. S. Ward (1985). *Poly Dimethylsiloxane-poly(e-Caprolactone) block copolymers: Synthesis and applications,* Polymer Preprints 26(1):251-252.

[147] Wang, Y., and M. A. Hillmyer. (2001). *Polyethylene-poly(L-lactide) diblock copolymers: synthesis and compatibilization of poly(L-lactide)/polyethylene blends,* Journal of Polymer Science, Part A: Polymer Chemistry 39:2755-2766.

[148] Frick, E. M., and M. A. Hillmyer (2000). Synthesis and characterization of polylactide-block-polyisoprene-block-polylactide triblock copolymers: New thermoplastic elastomers containing biodegradable segments, Macromolecular Rapid Communications 21(18): 13 17-1322.

[149] Frick, E. M., and M. A. Hillmyer (2001). *Polylactide-b-polyisoprene-b-polylactide triblock copolymers: New thermoplastic elastomers containing biodegradable segments,* Polymer Preprints 42(1):534.

[150] Mecerreyes, D., P. Dubois, R. Jerome, J. L. Hedrick, and C. J. Hawker (1999). *Synthesis of dendritic-linear block copolymers by living ring-opening polymerization of lactones and lactides using dendritic initiators,* Journal of Polymer Science, Part A: Polymer Chemistry 37:1923-1930.

[151] Kim, S. H., Y.-K. Han, Y. H. Kim, and S. I. Hong (1992). *Multifunctional initiation of lactide polymerization by stannous octoate/pentaerythritol,* Makromol Chem 193:1623-1631.

[152] Kim, S. H., Y.-K. Han, K.-D. Ahn, Y. H. Kim, and Y. Chang (1993). *Preparation of star-shaped polylactide with pentaerythritol and stannous octoate,* Makromol. Chem. 194:3229-3236.

[153] Lee, S.-H., S. H. Kim, Y.-K. Han, and Y.-H. Kim (2001). *Synthesis and degradation of end-group-functionalized polylactide,* Journal of Polymer Science, Part A: Polymer Chemistry 39:973-985.

[154] Arvanitoyannis, I., A. Nakayama, N. Kawasaki, and N. Yamamoto (1995). *Novel star-shaped polylactide with glycerol using stannous octoate or tetraphenyltin as catalyst: 1. Synthesis, characterization and study of their biodegradability,* Polymer 36(15) :2947-295 6.

[155] Trollsås, M., J. L. Hedrick, D. Mecerreyes, P. Dubois, R. Jérôme, H. Ihre, and A. Hult (1997). *Versatile and Controlled Synthesis of Star and Branched Macromolecules by Dentritic Initiation,* Macromolecules 30(26): 8508-8511.

[156] Trollsås, M., J. L. Hedrick, D. Mecerreyes, P. Dubois, R. Jérôme, H. Ihre, and A. Hult (1998). *Highly functional branched and dendri-graft aliphatic polyesters through ring opening polymerization,* Macromolecules 3 1(9):2756-2763.

[157] Breitenbach, A., and T. Kissel (1998). Biodegradable comb polyesters: Part 1 synthesis, characterization and structural analysis of poly(lactide) and poly (lactide-co-glycolide) grafted onto water-soluble poly(vinylalcohol) as backbone, Polymer 39(14): 3261-3271.

[158] Breitenbach, A., K. F. Pistel, and T. Kissel (2000). *Biodegradable comb polyesters. Part II. Erosion and release properties of poly(vinyl alcohol) -g-poly (lactide-co-glycolic acid),* Polymer 41:4781-4792.

[159] Kim, J.-B., C. Wang, M. L. Bruening, and G. L. Baker (2001). *Synthesis of poly(2-hydroxyethyl methacrylate)-block-poly(Lactide) on gold by dual living polymerizations,* Polymer Preprints 42(1):551-552.

[160] Li, Y., J. Nothnagel, and T. Kissel (1997). *Biodegradable brush-like graft polymers from Poly(D,L-lactide) or Poly(D,L-lactide-co-glycolide) and charge-modified, hydrophilic dextrans as backbone- sythesis, characterization and in vitro degradation properties*, Polymer 38(25):6197-6206.

[161] Donabedian, D. H., and S. P. McCarthy (1998). *Acylation of pullulan by ring-opening of lactones*, Macromolecules 31:1032-1039.

[162] Wallach, J. A., and S. J. Huang (2000). *Copolymers of itaconic anhydride and methacrylate-terminated poly (lactic acid) macromonomers*, Biomacromolecules 1:174-179.

[163] Lim, D. W., S. H. Choi, and T. G. Park (2000). *A new class of biodegradable hydrogels stereocomplexed by enantiomeric oligo(lactide) side chains of poly(HEMA-g-OLA)s*, Macromolecular Rapid Communications 21 (8):464-471.

[164] Trollsås, M., C. J. Hawker, J. F. Remenar, J. L. Hedrick, M. Johansson, H. Ihre, and A. Hult (1998). *Highly branched radial block copolymers via dendritic initaition of aliphatic aolyesters*, Journal of Polymer Science, Part A: Polymer Chemistry 36(1 5):2793-2798.

[165] Trollsås, M., J. Hedrick, D. Mecerreyes, R. Jérôme, and P. Dubois (1998). *Internal functionalization in hyperbranched polyesters*, Journal of Polymer Science, Part A: Polymer Chemistry 36(1 7):3 187-3192.

[166] Hedrick, J. L., M. Trollsås, C. J. Hawker, B. Atthoff, H. Claesson, A. Heise, and R. D. Miller (1998). *Dendrimer-like star block and amphiphilic copolymers by combination of ring opening and atom transfer radical polymerization*, Macromolecules 31(25): 8691- 8705.

[167] Carnahan, M. A., and M. W. Grinstaff (2001). *Synthesis and characterization of polyether-ester dendrimers from glycerol and lactic acid*, Journal of the American Chemical Society 123:2905-2906.

[168] Grijpma, D. W., C. A. P. Joziasse, and A. J. Pennings (1993). *Star-shaped polylactide-containing block copolymers*, Die Makromoleculare Chemie, Rapid Communications 147:155-161.

[169] Garlotta, D. (2001). *A literature review of poly(lactic acid)*, J Polym Environ 9(2):63-84.

[170] Kratochvil, P., and U. W. Suter (1988). *Definitions of terms relating to individual macromolecules, their assemblies, and dilute polymer solutions*, Pure Appl Chem 61:211-241.

[171] Ren, J., O. Urakawa, and K. Adachi (2003). *Dielectric study on dynamics and conformation of poly(D,L-lactic acid) in dilute and semi-dilute solutions*, Polymer 44(3):847-855.

[172] Schindler, A., and D. Harper (1979). *Polylactide II. Viscosity-molecular weight relationships and unperturbed chain dimensions*, J Polym Sci, Polym Chem Ed 17:2593-2599.

[173] Joziasse, C. A. P., H. Veenstra, D. W. Grijpma, and A. J. Pennings (1996). *On the chain stiffness of poly(lactide)s*, Macromol Chem Phys 197:2219-2229.

[174] Kang, S., G. Zhang, K. Aou, R. L. Pekrul, S. L. Hsu, and X. Yang (2002). *Raman and light scattering analysis of poly(lactic acid) flexibility*, Polym Preprints 43(2):896-897.

[175] Ren, J., O. Urukawa, and K. Adachi (2003). *Dielectric and viscoelastic studies of segmental and normal mode relaxations in undiluted poly (d,l-lactic acid),* Macromolecules 36:210-219.

[176] Dorgan, J. R., J. Janzen, S. B. Hait, and D. K. Knauss (2005). *Fundamental solution and single chain properties of polylactides,* J. Polym. Sci., Part B: Polym Phys. 43:3 100- 3111.

[177] Blomqvist, J. (2001). *RIS Metropolis Monte Carlo studies of poly[L-lactic), poly(L,D-lactic) and polyglycolic acids,* Polymer 42:3515-3521.

[178] Biresaw, G., and C. J. Carriere (2002). *Interfacial tension of poly(lactic acid)/polystyrene blends,* J. Polym. Sci., Part B: Polym Phys 40:2248-225 8.

[179] Cooper-White, J. J., and M. E. Mackay (1999). *Rheological properties of poly(lactides). Effect of molecular weight and temperature on the viscoelasticity of poly(l-lactic acid),* J. Polym. Sci., Part B: Polym Phys 37:1803-1814.

[180] Dorgan, J. R., and J. S. Williams (1999). *Melt rheology of poly(lactic acid): Entanglement and chain architecture effects,* Journal of Rheology 43(5):1 141-1155.

[181] Lehermeier, H. J., and J. R. Dorgan (2001). *Melt rheology of poly(lactic acid): Consequences of blending chain architectures,* Polymer and Engineering Science 41(12):2172-2184.

[182] Palade, L. I., H. J. Lehermeier, and J. R. Dorgan (2001). *Melt rheology of high-L content poly (lactic acid),* Macromolecules 34(5): 1384-1390.

[183] Dorgan, J. R., J. Janzen, M. P. Clayton, D. M. Knauss, and S. B. Hait (2005). *Melt Rheology of Variable L-content Poly (lactic acid),* Journal of Rheology 49(3): 607-619.

[184] Lehermeier, H. J. Poly (lactic acid): An environmentally benign polymer derived from corn: Rheological and permeation properties. M. S., Colorado School of Mines, Golden, Colorado, 2000.

[185] Leenslag, J. W. G., S., and A. J. Pennings (1984). Resorbable materials of poly(L-lactide). V. *Influence of secondary structure on the mechanical properties and hydrolyzability of poly(L-lactide) fibers produced by a dry-spinning method,* J Appl Polym Sci 29:2829-2842.

[186] Leenslag, J. W., and A. J. Pennings (1987). *High-strength fibers by hot drawing of as-polymerized poly(L-lactide),* Polymer Communications 28 (April):92-94.

[187] Fambri, L. P., A., R. Fenner, S. D. Incardona, and C. Migliaresi (1997). *Biodegradable fibres of poly(L-lactic acid) produced by melt spinning,* Polymer 38(1):79-85.

[188] Incardona, S. D. F., L., and C. Migliaresi (1996). *Poly-L-lactic acid braided fibres produced by melt spinning: characterization and in vitro degradation,* J Mater Sci Mater Med 7:387-391

[189] Schmack, G. T., B., R. Vogel, R. Beyreuther, S. Jacobsen, and H. G. Fritz (1999). *Biodegradable fibers of poly(L-lactide) produced by high-speed melt spinning and spin drawing,* J Appl Polym Sci 73:2785-2797.

[190] Mezghani, K. S., and J.E. (1998). *High speed melt spinning of poly(L-lactic acid) filaments,* J Polym Sci Part B: Polym Phys 36:1005-1012.

[191] Hoogsteen, W., A. R. Postema, and A. J. Pennings (1990). *Crystal structure, conformation, and morphology of solution-spun poly(L-lactide) fibers,* Macromolecules 23(2):634-642.

[192] Cicero, J. A., J. R. Dorgan, J. Janzen, J. Garrett, J. Runt, and J. S. Lin (2002). *Supramolecular morphology of two-step melt-spun poly (lactic acid) fibers,* J Appld Polym Sc 86:2828-2838.

[193] Cicero, J. A., J. R. Dorgan, J. Garrett, J. Runt, and J. S. Lin (2002). *Effects of molecular architecture on two-step melt-spun poly (lactic acid) fibers,* J Appld Polym Sci 86:283 9- 2846.

[194] Cicero, J. A., J. R. Dorgan, S. F. Dec, and D. M. Knauss (2002). *Phosphite stabilization effects on two-step melt-spun fibers of polylactide,* Polymer Degradation and Stability 78(1):95-105.

[195] Siparsky, G. L., K. J. Voorhees, J. R. Dorgan, and K. Schilling (1997). *Water transport in poly (lactic acid), poly (lactic acid-co-caprolactone) copolymers, and poly (lactic acid)/Polyethylene glycol blends,* Journal of Environmental Polymer Degradation 5(3): 125-136.

[196] Lehermeier, H. J., J. R. Dorgan, and J. D. Way (2001). *Gas permeation properties of poly (lactic acid),* J Memb Sci 190:243-25 1.

[197] Hauschild, M. (2005). *Assessing environmental impacts in a life-cycle perspective,* Environ Sci Technol 39:81A-88A.

[198] Utraki, L. A., and K. C. Cole (2002). *First International Symposium on Polymeric Nanocomposites,* Polym Eng Sci 1799-1937.

[199] Pluta, M., A. Galeski, M. Alexandre, M.-A. Paul, and P. Dubois (2002). *Polylactide/montmorillonite nanocomposites and microcomposites prepared by melt blending: structure and some physical properties,* J Appl Polym Sci 86:1497-1506.

[200] Angeles, M. N., and A. Dufresne (2000*). Plasticized starch/tunicin whiskers nanocomposite materials,* Macromolecules 33:8344.

[201] Angeles, M. N., and A. Dufresne (2001). *Plasticized starch/sunicin whiskers nanocomposite materials.* 2. Mechanical Properties, Macromolecules 34.

[202] Cavaille, J. Y., H. Chanzy, V. Favier, and B. Ernst (2000). Cellulose microfibril-reinforced polymers and their applications. U.S. Patent 6,103,790.

[203] Cavaille, J. Y., H. Chanzy, E. Fleury, and J. F. Sassi (2000). Surface-modified cellulose micro fibrils, method for making the same, and use thereof as filler in composite materials. U.S. Patent 6,117,545

[204] Dufresne, A. (1998). *High performance nanocomposite materials from thermoplastic matrix and polysaccharide fillers, Recent Research Developments in Macromolecules* Research 3(Pt. 2):455-474.

[205] Dufresne, A., J.-Y. Cavaille, and W. Helbert (1996). *New nanocomposite materials: microcrystalline starch reinforced thermoplastic,* Macromolecules 29:7624-7626.

[206] Dufresne, A., M. B. Kellerhals, and B. Witholt (1999). *Transcrystallization in mclPHA s/Cellulose whisker composites,* Macromolecules 32:7396-7401.

[207] Teramoto, Y., and Y. Nishio (2003*). Cellulose diacetate-graft-poly (lactic acid)s: synthesis of wide-ranging compositions and their thermal and mechanical properties,* Polymer 44:2701-2709.

[208] Ritter, S. K. (2004). Biomass or bust, Chemical and Engineering News(May 31):31-34.

[209] Minoru, N. (1999). *Preparation and characterization of novel biodegradable optically active polyesters from malic acid,* Macromolecules 32:7762-7767.

[210] Nagata, M., and M. Nakae (2001). *Synthesis, characterization, and in vitro degredation of therm otropic polyesters and copolyesters based on terephthalic Acid, 3-(4-*

hydroxyphenyl)propionic acid, and glycols, Journal of Polymer Science, Part A: Polymer Chemistry 39:3043-305 1.

[211] Lubbert, A., and S. B. Jorgenson (2001). *Bioreactor performance: A more scientific approach for practice,* J Biotechnol 85:187-212.

[212] Gruber, P. R., M. H. Hartmann, J. J. Kolstad, D. R. Witzke, and A. L. Brosch (1996). *Method of crosslinking PLA.* PCT 94/08 508.

[213] Gruber, P. R., J. J. Kolstad, M. H. Hartmann, and A. L. Brosch (1997). *Viscosity-modified lactide polymer composition and process for manufacture thereof.* U.S. Patent 5,594,095, Jan. 14.

[214] Kanda, T., K. Onishi, Y. Kichise, and Y. Fujii (1997). Aromatic polyester-polyoxyalkylene block copolymers polymerized with lactic acid and their preparation methods. 09100345.

[215] O'Brien, W. G., L. A. Cariello, and T. F. Wells (1996). *Integrated process for the manufacture of lactide.* U.S. Patent 5,521,278, May 28.

C. POLYHYDROXYALKANOATES

1. Introduction

A variety of bacteria synthesize PHAs as intracellular carbon and energy storage materials, much as higher organisms synthesize starch, glycogen, or fat for energy storage. PHAs were first observed as dark-staining bodies within microbes observed under microscopes, and in 1925, poly(3HB), was first isolated from Bacillus megaterium by Lemoigne [1, 2]. Because they are bacterial storage polymers, PHAs are of necessity biodegradable, and further study revealing their plasticity spurred commercial interest. In 1974, Wallen and Rohwedder predicted that novel hydroxyalkanoate (HA) subunits could become important to further development of the microbial polyester, due to the ability of varying structural units within poly(3HB) to modify polymer properties [3].

Research investigating the identities and characteristics of various units within microbial PHAs began in earnest during the 1980s, when a number of bacteria were found to synthesize optically active homopolymers and copolymers of (R)-3-hydroxyalkanoates [(R)-3HAs] ranging from 4 to 14 carbon atoms [4-9]. Saturated, unsaturated, halogenated, branched, and aromatic side chains in (R)-3HA monomeric units have now been found within microbial PHAs; at present, at least 140 different monomeric units have been found as constituents of microbial PHAs [10-12]. In addition, certain bacteria also produce copolymers containing hydroxyalkanoate repeat units with side chains such as 3-hydroxypropionate and 4-hydroxybutyrate [13, 14]. Subsequently, many different approaches have been taken to produce PHA materials with desirable physical properties [15-17].

2. PHA Biosynthesis, Biodegradation, and Environmental Impact: Overview

The biosynthetic routes to PHA monomer synthesis are interconnected with several central metabolic pathways: the tricarboxylic acid (TCA) cycle, fatty acid degradation (β-

oxidation), and lipid biosynthesis, through common intermediates and cofactors such as acetyl-CoA, other fatty acyl CoAs, NADH, and NADPH.

PHA is synthesized under the direction of a family of enzymes that shows some diversity among microorganisms but nevertheless possesses several conserved members. β-ketothiolases harvest acetyl CoA from the TCA cycle, condensing it into acetoacetyl CoA; transacylases similarly harvest intermediates from lipid biosynthesis, and enoyl CoA hydratases divert enoyl CoA intermediates from β-oxidation. Where necessary, as with acetoacetyl CoA, reductases then reduce keto groups to hydroxy groups. Each of these enzymes thus provides monomers for PHA synthesis by the final enzyme in the pathway, PHA synthase. The β-ketothiolases, transacylases, hydratases, reductases, and PHA synthases are encoded by well-understood genes: PhaA, PhaG, PhaJ, PhaB, and PhaC, respectively. The genetic understanding of this system, as well as its connectedness to other well-understood metabolic pathways, has made PHA biosynthesis an ideal target for metabolic engineering.

Once the PHA is synthesized, the microbes are harvested, disrupted, and fractionated to isolate the polyester. To produce PHA materials with desirable physical properties, a variety of approaches including copolymerization, blending with other polymers, cross-linking, and the introduction of functional groups have been studied extensively. PHAs can be produced in a wide variety of molecular configurations; homopolymers, copolymers, and functionalized polymer chains may all be created by utilizing various microorganisms and fermentation conditions. A representative sample is shown in Figure 12 [18].

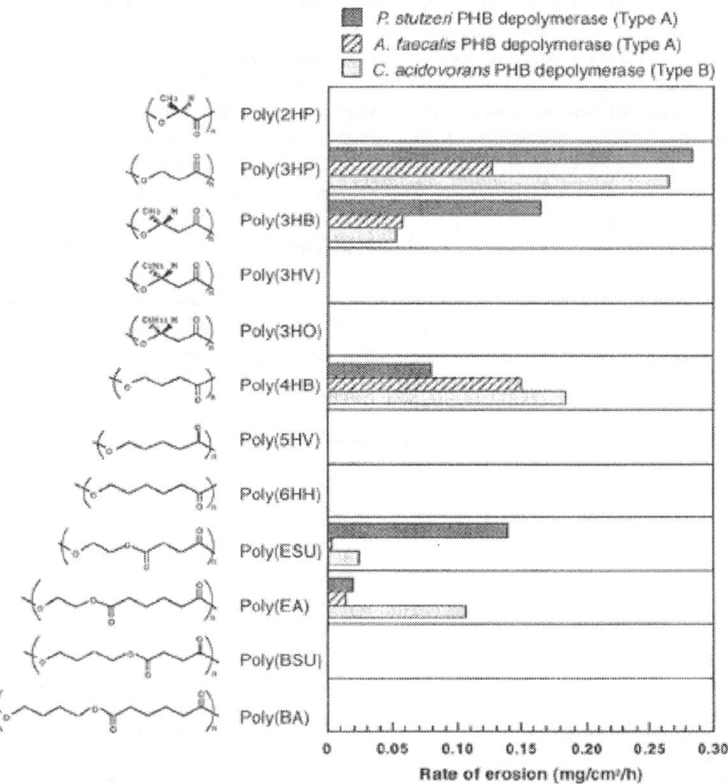

Figure 12. Chemical structures of PHAs and degradation rates in aqueous solutions at 37 °C containing depolymerases [18].

PHAs are benign materials from an environmental perspective for several reasons. First, substrates necessary for bacterial growth of PHA-producing bacteria are available from renewable resources, and second, PHAs are highly biodegradable, not only by their bacterial producers, but also by numerous other aquatic and terrestrial microorganisms. Chowdhury first isolated poly(3HB)-degrading bacteria in 1963 [19], and additional studies on the isolation and characterization of PHA-degrading microorganisms began to appear in the 1990s [18]. Since that time, concern about the environmental impacts of plastic wastes has led to expanded investigations into PHA-degrading microorganisms. At this point, a number of extracellular PHA-degrading enzymes from various microorganisms have been purified and characterized, and PHA-derived products appear to decompose readily in both composting and activated sludge systems [20, 21].

3. State of the Science

3.1. Synthesis

Poly (3HB) is the most common biological polyester and is produced by numerous microorganisms in nature [22]. From this basic starting point, however, three primary approaches have been investigated to produce novel polymers with a wide range of properties.

3.3.1. Feedstock manipulation. Because the physical and thermal properties of PHA polymers and copolymers can be regulated by varying their molecular structure and copolymer compositions, the simplest metabolic engineering strategy is to provide specific carbon sources to the microbes that bias the monomer production in favor of desired compounds. Polymer composition can be further controlled by varying the feed ratio of various substrates.

For example, a random copolymer of (R)-3HB and (R)-3 -hydroxyvalerate, poly(3HB-co-3HV), has been produced in Ralstonia eutropha by feeding pentanoic acid and butyric acid as the carbon sources [5]. The poly(3HB) homopolymer was produced from butyric acid, while a poly(3HB-co-3HV) copolymer was produced from pentanoic acid. By varying the ratio of pentanoic acid to butyric acid in the feed, variable composition copolymers were produced. Similarly, using 3-hydroxypropionic acid as the substrate, R. eutropha produced a random copolyester of (R)-3HB and 3-hydroxypropionate, poly(3HB-co-3HP) [23]. The same copolymer can be produced by Alcaligenes latus [24]. Using olive oil as a substrate, Aeromonas caviae produced a random copolymer of (R)-3HB and (R)-3-hydroxyhexanoate, poly(3HB-co3HHx) [25]. Feeding 4-hydroxybutyric acid, 1,4-butanediol, or butyrolactone as the carbon source produces a random copolyester of (R)-3HB and 4-hydroxybutyrate, poly(3HB-co-4HB) when R. eutropha, [5] A. latus, [26] or Comamonas acidovorans [27] are utilized. Recently, a number of unusual sulfur-containing polymers have been generated by feeding alkylthionates to R. eutropha, and external substrate manipulation has even been used to generate block copolymers by intermittent addition of one substrate. In the latter example, R. eutropha was fed pulses of valerate–a precursor substrate of 3 -hydroxyvalerate–to medium containing an excess of fructose–a precursor of PHB–to generate the block copolymer [12].

PHAs with functional groups in the side chains can also be produced when functionalized organic substrates are employed, allowing the engineering of specific physical properties and the provision of reactive sites for applications such as adhesives and coatings. Representative side chains that have been incorporated into 3HA include unsaturated [8, 28], halogenated [10], branched [29], and aromatic [30] moieties.

The addition of inhibitors, reducing the function of metabolic pathways that compete with PHA synthesis, is another feeding strategy that has been employed to improve PHA yield and/or to facilitate incorporation of longer monomers. In one example, acrylate was used to inhibit β-oxidation in R. eutropha, such that the microbe accumulated a copolymer of both short- and medium-chain length monomers rather than exclusively short chain-length monomers [12].

3.1.2. Genetic engineering. Extensive investigations into PHA synthesis has led to the identification, cloning, and sequencing of approximately 40 responsible genes from various gram-positive and gram-negative bacteria, providing a great deal of material to support efforts in genetic engineering [31].

Among the most commonly-used heterologous hosts, including Ralstonia eutropha, Pseudomonas putida, Pseudomonas oleovorans, and Escherichia coli, the latter deserves special mention for its advantages as a host. It has an extremely thoroughly understood physiology; it is not a native PHA producer, with the result that productivity is not limited by natural regulation; it harbors no native machinery for PHA degradation; and its cells are easily disrupted, facilitating PHA recovery [12].

One straightforward approach to increasing PHA synthesis is simply increasing gene dosage or providing additional copies of the PHA synthetic enzymes. While this has shown success in efforts with Aeromonas punctata and R. eutropha, not all attempts have been successful. This shows that the effectiveness of gene dosage depends on the limiting factor for polymer synthesis in each individual case.

A more involved but increasingly popular approach is the expression of heterologous genes for polymer precursor production in a desirable host, with the goals of facilitating synthesis of polymers that would not naturally accumulate and/or that might have desirable structures and properties, as well as facilitating use of simple, inexpensive carbon sources for production of the desired polymers. In one encouraging example, the incorporation of three enzymes into a recombinant Salmonella allowed the synthesis of propionate, an expensive but previously necessary substrate for the synthesis of PHB-co-PHV, from succinyl-CoA—an intermediate in the TCA cycle. The microbe was then able to synthesize PHB-co-PHV from glycerol, a significantly less expensive carbon source. A number of similar efforts have also been successful [12].

In the realm of pathway engineering, competing pathways can also be eliminated and regulatory systems can be altered to facilitate PHA synthesis. Related to the example above involving propionate, propionate-degrading enzymes have been deleted, and propionate synthesis machinery has been placed under the control of an IPTG-inducible promoter, allowing the composition of PHB-co-PHV polymers to be adjusted at will [12].

Finally, PHA biosynthetic enzymes are amenable to protein engineering, and PHA synthase in particular is the subject of an on-going effort to develop a complete understanding of its structural and mechanical characteristics. In the case of the Pseudomonas 61-3 PHA synthase, error-prone PCR mutagenesis revealed two primary sites that affected PHB accumulation, allowing subsequent site-directed mutagenesis to test all possible amino acid combinations at those sites. The optimal combination of amino acids at those sites yielded a synthase that promoted the accumulation of up to 400-fold more PHB in the microbe [12].

Given the number and diversity of PHA synthase genes now available, family shuffling approaches are being explored as well, and transacylase and hydratase genes are expected to be targets in the near future [12].

Mathematical modeling, informed by microarray analysis, proteomics, and metabolomics, is also suggesting new targets for additional metabolic engineering efforts. Flux analysis, for example, recently elucidated the role of the Entner-Doudoroff pathway in PHB production in E. coli, while other mathematical models have identified optimal substrate switching strategies for the production of desired block copolymers [12].

3.1.3. Transgenic plants. The detailed genetic understanding of PHA biosynthesis pathways offers hope for the cost-effective production of PHAs in transgenic plants [32, 33]. Initial success in Arabidopsis thaliana suffered from a depletion of suitable substrate for growth [34], but genetic manipulation led to more effective production in plant plasmids [35]. The construction of transgenic A. thaliana using the PHA syntheses gene of P. aeruginosa indicates that plant fatty acids can generate a range of (R-)3-hydroxyalkanoate monomers that can be used to synthesize medium molecular weight PHAs [36]. From a commercial perspective, the copolymers P(3HB-co-3HV) and P(3HB-co-4HB) are attractive, and recent reports show promise for the production of both monomers and polymers plants [16, 14].

Development of PHA-producing transgenic switchgrass is also underway, with the goal of incorporating the synthesis into a more productive, easily-grown plant [37].

In 2001, Metabolix Inc. was awarded a $15 million "Industries of the Future" cost-shared grant from DOE to help fund the development of a biomass biorefinery based on switchgrass. The goal of this program is to produce PHAs in plants and, after polymer extraction, use the residual plant biomass for fuel generation, thereby generating both materials and fuels from a sustainable resource (http://www.metabolix.com/biotechnologypercent20foundation /plants.html). Efforts are also underway to produce PHAs/PHBs in photosynthetic bacteria.

3.2. Physical Properties

The molecular weight (MW) of poly(3HB) produced from wild-type (unmodified) bacteria usually ranges from 104 to 106 grams per mole, with a polydispersity of about 2 [38]. Within the bacteria, the polymers remain amorphous and are found as water-insoluble inclusions [39, 40]. This fact is surprising, as the polyester only has (R)-configuration stereochemical sites in the backbone and thus exhibits a perfectly isotactic structure. Crystallinity is typically 55-80 percent in poly(3HB) isolated from bacteria [7].

3.2.1. Homopolymers. The physical properties of amorphous, crystalline, and ultrahigh MW poly(3HB) are tabulated below, showing comparisons with isotactic PP (Table 1) [41, 9, 42]. The primary difficulty with ordinary poly(3HB) is that it is a relatively brittle plastic, as shown by the low extension to break value in comparison to polypropylene [43].

Films have been prepared from ultra-high molecular-weight poly(3HB) that show improved mechanical properties when stretched, however, raising both the elongation to break and tensile strength values [44, 45]. Mechanical properties have been improved further upon annealing, as well as by the incorporation of structural variations made possible by genetic engineering, with the result that new PHAs are promising candidates for further commercial exploitation.

Table 1. Physical properties of PHA homopolymers

Plastic	Density g/cm	Tg°C	Tm°C	Crystal structure	Conformation	Morphology	Young's modulus GPa	Tensile strength MPa	Extension to break percent
poly(3HB), amorphous	1.18	4	180		left-handed 21 helix		3.5	43	5
poly(3HB), crystalline	1.26			ortho-rhombic		lamellar			
ultra-high M_w poly(3HB)						lamellar	1.1	62	58
PP									400

3.2.2. Copolymers. Random copolymers of (R)-3HB and (R)-3HV have been investigated extensively with regard to various physical properties. In the case of random copolymers of (R)-3HB with (R)-3HHX (3-hydroxyhexanoate), the repeat unit contains a propyl side chain that strongly affects properties. Most notably, in solution-cast films, poly(3HB-co-3HHx) becomes soft and flexible with increased proportion of (R)-3HHx, as shown by the diminished Tm, Tg, crystallinity, and tensile strength, and by the increased elongation to break factor (Table 2). An additional example is provided by poly(3HB-co-4HB), in which crystallinity, elongation to break, and other physical properties can vary widely by adjusting proportions of the component monomers (Table 2). These data show that PHA copolymer properties, like those of other plastics, can be regulated by their composition; in addition, the range of materials observed, from hard crystalline plastics to elastic rubbers, show that these plastics hold great promise for a wide variety of applications.

Table 2. Physical properties of PHA copolymers

Plastic	non-BIB mole fraction percent	Crystallinity percent	Tm°C	Tg °C	Tensile strength MPa	Elongation to break percent
poly(3HB-co-3HV)[a]	various	50-70				
poly(3HB-co-3HHx)[b]	0	60	177	4	43	6
	17				20	850
.	25	18	52	-4		
poly(3HB-co-4HB)[c]	0	60			43	5
	16				26	444
	49	14				
	64				17	
	100				104	

a[46]
b[47]
c[27]

3.2.3. Blends. An established method for changing plastic properties is through blending with other materials. Several detailed investigations of blends containing poly(3HB) with other biodegradable polymers have appeared. Such blends are physical mixtures of different polymers; however, sometimes the two polymers react with one another, compatibilizing the blend. Blends can be either homogeneous, forming a single thermodynamic phase, or heterogeneous, comprising two or more phases. Blend properties are dependent, in turn, on such phase behavior.

Miscible blends containing poly(3HB) have been formed with poly(ethylene oxide) [48, 11, 49], poly(vinyl alcohol), [50] PLA [51, 52], poly(ε-caprolactone-co-lactide) [53], poly(butylene succinate-co-butylene adipate) and poly(butylene succinate-co-ε-caprolactone) [54]. Immiscible blends are formed in mixtures of poly(3HB) with poly(β-propiolactone)[55, 13], poly(ethylene adipate) [55], poly(butylene adipate) [56], and poly(ε-caprolactone) [55].

For miscible blends of poly(3HB) with atactic poly(3-hydroxybutyrate), increasing the weight content of the latter from 0–76 percent increases the elongation to break from 5–500 percent with an accompanying decrease in the Young's modulus and tensile strength [15]. For the immiscible system of poly(ε-caprolactone) with poly(3HB), in contrast, the decrease in Young's modulus and tensile strength is not accompanied by an increase in elongation to break due to macroscopic phase separation [56]. These studies show the continuing importance of physical property modification by blending as a valuable route towards improving the properties and therefore increasing the utilization of bioplastics.

4. Research Priorities

PHA development is proceeding in a number of promising directions on both metabolic engineering and chemical engineering fronts. Fortunately, most of these have the potential for success both individually and in combination with others, such that no particular obstacle is currently forming a bottleneck to further progress. The range of physical and thermal properties achievable with PHAs is still expanding rapidly as new configurations of copolymers and blends are explored. An on-going challenge will be the ability of the metabolic engineers to keep pace with the discoveries of the materials scientists, enabling microbes to synthesize the desired polymers both conveniently and inexpensively. These efforts can be categorized as follows.

4.1. Investigation of Novel Polymers and Properties

Clearly, a number of modifications of PHA composition have the potential to improve the plasticity, moldability, heat tolerance, and durability of the resulting plastics to approach those of conventional thermoplastics. Because of the promising availability and flexibility of routes to PHA synthesis, and because of increasing oil prices that will enable PHA polymers to become increasingly cost-competitive, it is a valuable effort to explore the properties of new PHA-based homopolymers, copolymers, and blends even before microbial pathways to their syntheses are in place.

4.2. Metabolic and Genetic Engineering

The increasing availability of mathematical modeling tools, genomic and proteomic data and techniques, and microarray and antisense RNA technology will allow increasingly accurate prediction of useful targets for metabolic engineering. At the same time, genetic manipulation within both microorganisms and plants is becoming increasingly possible and rapid. Several enzymes central to PHA synthesis are just beginning to be explored through combinatorial and rational design mutagenesis approaches, and efforts to understand their catalytic mechanisms, substrate specificities, modes of competition with other enzymes, and regulation are likely to contribute greatly to the microbial or plant-based syntheses of novel polymers.

4.3. Reactor and Processing Technology

As described previously, gains in commercial feasibility are often found in improving bioreactor yield and in diminishing processing costs. Without reiterating topics addressed previously, it is nevertheless important to include these in the research priorities for PHA efforts, with special note that the transfer of PHA synthesis to plants may circumvent many limitations of both reactor efficiency and processing costs.

5. Commercialization

PHA-based materials were originally produced primarily under the trade name of Biopol[TM] by ICI, Zeneca, and Monsanto. Metabolix recently acquired the Biopol[TM] patents from Monsanto, however, and is now producing several PHA polymers in bacteria and plants. Independently, Proctor and Gamble developed PHA copolymers of short and medium chain-length monomers under the trade name NodaxTM and licensed the technology to the Kaneka Corporation, a Japanese manufacturer of plastics and resins that is focusing on the production of P(3HB-co-3HHX).

Nevertheless, Metabolix, Inc. currently possesses a virtual monopoly on the commercial research and development of PHA polymers, holding approximately 90 issued U.S. patents and approximately 40 additional pending U.S. patents, as well as their foreign equivalents, protecting methods of PHA isolation, purification, and processing; use of several preferred metabolic pathways for PHA copolymer production; and several novel PHA compositions and specific applications. Most basic to Metabolix's position are the patents that give Metabolix exclusive rights to the genes within the PHA biosynthesis pathway, as well as the use of the genes in any combination for the preparation of PHAs (e.g., [57, 58]). As the company states itself on its Web site, "Metabolix owns the genes that encode the basic PHA pathway." (http://www.metabolix.com/publications/patents.html).

This comprehensive ownership of intellectual property has given Metabolix the freedom and protection to invest in research to improve PHA biosynthesis and processing technologies, with which it has greatly advanced these fields, as evident from the preceding discussion. At the same time, the present situation effectively diminishes or even deprives potential competitors of the ability to use either microbes or plants to produce PHAs commercially.

REFERENCES

[1]   Lemoigne, M. (1926). *Produit de deshydratation et de polymerisation de l'acide beta-oxybutyrique,* Bull Soc Chim Biol 8:770-782.

[2]   Lemoigne, M. (1927). *Etudes sur l-autolyse microbienne origigine de l'acide beta-oxybutyrique forme par autolyse,* Ann Inst Pasteur 41:148-165.

[3]   Wallen, L. L., and W. K. Rohwedder (1974). *Poly-beta-hydroxyalkanoate from activated sludge,* Environ Sci Technol 8:576-579.

[4]   Barham, P. J., A. Keller, E. L. Otun, and P. J. Holmes (1984). *Crystallization and morphology of a bacterial thermoplastic: poly-3-hydroxybutyrate,* J Mater Sci 19:278 1- 2794.

[5]   Doi, Y., M. Kunioka, Y. Nakamura, and K. Soga (1988*). Nuclear magnetic resonance studies on unusual bacterial copolyesters of 3-hydroxybutyrate and 4-hydroxybutyrate,* Macromolecules 21:2722-2727.

[6]   Doi, Y., A. Tamaki, M. Kunioka, and K. Soga (1988). *Production of copolyesters of 3-hydroxybutyrate and 3-hydroxyvalerate by Alcaligenes eutrophus from butyric and pentanoic acids,* Appl Microbiol Biotechnol 28:330-334.

[7]   Holmes, P. A. (1988). *Biologically Produced (R)-3-hydroxyalkanoate polymers and copolymers, in Developments in Crystalline Polymers* (D. C. Bassett, Ed.). Elsevier, London, pp 1-65.

[8]   Lageveen, R. G., G. W. Huisman, H. Preusting, P. Ketelaar, G. Eggink, and B. Witholt (1988). *Formation of polyesters by Pseudomonas oleovorans: Effects of substrates on formation and composition of poly-(R)-3-hydroxyalkanoates and poly-(r)-3-hhydroxyalkenoates,* Appl Environ Microbiol 54:2924-2932.

[9]   Marchessault, R. H., T. L. Bluhm, Y. Desalandes, G. K. Hamer, W. J. Orts, P. R. Sundararejan, M. G. Taylor, S. Bloembergen, and D. A. Holden (1988). *Poly(b-hydroxyalkanoates): Biorefinery polymers in search of applications,* Macromol Chem Macomol Symp 19:235-254.

[10]   Abe, C., Y. Taima, Y. Nakamura, and Y. Doi (1990). *New bacterial copolyester of 3-hydroxyalkanoates and 3-hydroxy-w-fluoroalkanoates produced by Pseudomonas oleovorans,* Polym Comm 31:404-406.

[11]   Avella, M., E. Martuscelli, and P. Greco (1991). *Crystallization behavior of poly (ethylene oxide) from poly(3-hydroxybutyrate)/poly(ethylene oxide) blends: phase structuring, morphology and thermal behavior,* Polymer 32:1647-1653.

[12]   Aldor, I. S., and J. D. Keasling (2003). *Process design for microbial plastic factories: Metabolic engineering of polyhydroxyalkanoates,* Curr Opin Biotechnol 14:475-483.

[13]   Cao, A., N. Asakawa, N. Yoshie, and Y. Inoue (1998). *Phase structure and biodegradation of the bacterial poly(3-hydroxybutyric acid)/chemosythetic poly(3-hydroxypropionic acid) blend,* Polym J 30:743-752.

[14]   Valentin, H. E., S. E. Reiser, and K. J. Gruys (2000). *Poly(3-hydroxybutyrate-co-4-hydroxybutyrate) formation from g-aminobutyrate and glutamate,* Biotechnol Bioengng 67:291-299.

[15]   Kumagai, Y., and Y. Doi (1992). *Physical properties and biodegradability of blends of isotactic and atactic poly(3-hydroxybutyrate),* Makromol Chem Rapid Commun 13:179- 183.

[16] Slater, S., T. A. Mitsky, K. L. Houmiel, M. Hao, S. E. Reiser, N. B. Taylor, M. Tran, H. E. Valentin, D. J. Rodriguez, D. A. Stone, S. R. Padgette, G. Kishore, and K. J. Gruys (1999). *Metabolic engineering of Arabidopsis and Brassica for poly(3-hydroxybutyrate-co-3-hydroxyvalerate) copolymer production*, Nature Biotechnol 17:1011-1016.

[17] Nagata, M., and M. Nakae (2001). *Synthesis, characterization, and in vitro degradation of therm otropic polyesters and copolyesters based on terephthalic acid, 3-(4-Hydroxyphenyl)propionic acid, and glycols*, J Polym Sci A: Polym Chem 39:3043-3051.

[18] Doi, Y., K. Mukai, K. Kasuya, and K. Yamada (1994). *Biodegradation of biosynthetic and chemosynthetic polyhydroxyalkanoates*, in Biodegradable plastics and polymers (Y. Doi and K. Fukuda, Eds.). Elsevier, Amsterdam, pp 39-51.

[19] Chowdhury, A. A. (1963). *Poly(beta)-hydroxybuttersaureabbeauende Bakterien und Exoenzyme*, Archiv Mikrobiol 47:167-200.

[20] Elbanna, K., T. Lutke-Eversloh, D. Jendrossek, H. Luftmann, and A. Steinbuchel (2004). *Studies on the biodegradability of polythioester copolymers and homopolymers by polyhydroxyalkanoate (PHA)-degrading bacteria and PHA depolymerases*, Arch Microbiol 182:212-225.

[21] Wang, Y.-W., W. Mo, H. Yao, Q. Wu, J. Chen, and G.-Q. Chen (2004). *Biodegradation studies of poly(3-hydroxybutyrate-co-3-hydroxyhexanoate)*, Polym Deg Stabil 85:815-821.

[22] Dawes, E. A., and P. J. Senior (1973). *The role and regulation of energy reserve polymers in microorganisms*, Adv Microb Physiol 10:136-267.

[23] Nakamura, S., M. Kunioka, and Y. Doi (1991). *Biosynthesis and characterization of bacterial poly(3-hydroxybutyrate-co-4-hydroxproprionate)*, Macromol Rep (A) 28:15-24.

[24] Hiramatsu, M., and Y. Doi (1993). *Microbial synthesis and chracterization of poly(3-hydroxbutyrate-co-3-hydroxyproprionate)*, Polymer 34:4782-4786.

[25] Shimamura, E., K. Kasuya, G. Kobayashi, T. Shiotani, Y. Shima, and Y. Doi (1994). *Physical properties and biodegradability of microbial poly(3-hydroxybutyrate-co-3-hydroxyhexanoate)*, Macromolecules 27:878-880.

[26] Hiramatsu, M., N. Koyama, and Y. Doi (1993). *Production of poly(3-hydroxybutyrate-co-4-hydroxybutyrate) by Alcaligenes latus*, Biotechnol Lett 15:461-464.

[27] Saito, Y., and Y. Doi (1994). *Microbial Synthesis and properties of poly(3-hydroxybutyrate-co-4-hydroxbutyrate) in Comamonas acidovorans*, Biol Macromol 16:99-104.

[28] Fritzsche, K., R. W. Lenz, and R. C. Fuller (1990). *Production of unsaturated polyesters by Pseudomonas oleovorans*, Int J Biol Macromol 12:85-91.

[29] Fritzsche, K., R. W. Lenz, and R. C. Fuller (1990*). Bacterial polyesters containing branched poly (b-hydroxalkanoate) units*, Intl J Biol Macromol 12:92-101.

[30] Kim, Y. B., R. W. Lenz, and R. C. Fuller (1991). *Preparation and characterization of poly(b-hydroxyalkanoates) obtained from by Pseudomonas oleovorans grown with mixtures of 5-phenylvaleric acid and n-alkanoic acids*, Macromolecules 24:5256-5260.

[31] Rehm, B. H. A., and A. Steinbuchel (1999). *Biochemical and genetic analysis of PHA synthases and other proteins required for PHA synthesis*, Intl J Biol Macromol 25:3-19.

[32] Pool, R. (1989). *In search of the plastic potato*, Science 245:1187-1189.

[33] Van der Leij, F. R., and B. Witholt (1995). *Strategies for the sustainable production of new biodegradable polyesters in plants: A review*, Can J Microbiol(Suppl) 41:222-238.

[34] Poirier, Y., D. E. Dennis, K. Klomparens, and C. Somerville (1992). *Polyhydroxybutyrate, a biodegradable thermoplastic produced in transgenic plants*, Science 256:520-523.

[35] Nawrath, C., Y. Poirier, and C. Somerville (1994). *Targeting of the polyhydroxybutyrate biosynthetic pathway to the plasmids of Arabidopsis thaliana results in high levels of polymer accumulation*, Proc Natl Acad Sci USA 91:12760-12764.

[36] Mittendorf, V., E. J. Robertson, R. M. Leech, N. Kruger, A. Steinbuchel, and Y. Poirier (1998). *Synthesis of medium-chain-length polyhydroxyalkanoates in Arabidopsis thaliana using intermediates of peroxisomal fatty acid b-oxidatioin*, Proc Natl Acad Sci USA 95:13397-13402.

[37] Snell, K. D., and O. P. Peoples (2002). *Polyhydroxyalkanoate polymers and their production in transgenic plants*, Metab Eng 4:29-40.

[38] Doi, Y. (1990). *Microbial Polyesters*. VCH Publishers, New York.

[39] Barnard, G. N., and J. K. M. Sanders (1989). *The poly-b-hydroxybutyrate granule in vivo: a new insight based on NMR spectroscopy of whole cells.*, J Biol Chem 264:3286-3291.

[40] Amor, S. R., T. M. Rayment, and J. K. M. Sanders (1991). *Poly(hydroxybutyrate) in vivo: NMR and x-ray characterization of the elastomeric state*, Macromolecules 24:4583-4588.

[41] Lundgren, D. G., R. Alper, C. Schnaitman, and R. H. Marchessault (1965). *Characterization of poly-b-hydroxybutyrate extracted from different bacteria*, J Bacteriol 89:245-251.

[42] Iwata, T., and Y. Doi (1999). *Crystal structure and biodegradation of aliphatic polyester crystals*, Macromol Chem Phys 200:2429-2442.

[43] Khanna, S., and A. K. Srivastava (2005). *Recent advances in microbial polyhydroxyalkanoates*, Process Biochem 40:607-619.

[44] Kusaka, S., T. Iwata, and Y. Doi (1998). *Microbial synthesis and physical properties of ultrahigh molecular weight poly[(R)-3-hydroxybutyrate]*, J Macromol Sci Pure Appl Chem A35:319-335.

[45] Kusaka, S., T. Iwata, and Y. Doi (1999). *Properties and biodegradability of ultrahigh molecular weight poly[(R)-3-hydroxybutyrate] produced by recombinant Escheria coli*, Intl J Biol Macromol 25:87-94.

[46] Orts, W. J., R. H. Marchessault, and T. L. Bluhm (1991). *Thermodynamics of the melting point depression in poly(b-hydroxybutyrate-co-b-hydroxyvalerate) copolymers*, Macromolecules 24:6435-6438.

[47] Doi, Y., S. Kitamura, and H. Abe (1995). *Microbial synthesis and characterization of poly(3-hydroxybutyrate-co-3-hydroxyhexanoate)*, Macromolecules 28:4822-4828.

[48] Avella, M., and E. Martuscelli (1988). *Poly-D-(3-hydroxybutyrate)/poly(ethylene oxide) blends: phase diagram, thermal and crystallization behavior*, Polymer 29:1731-1737.

[49] Kumagai, Y., and Y. Doi (1992). *Enzymatic degradation of poly(3-hydroxybutyrate)-based blends: poly(3-hydroxybutyrate)/poly(ethylene oxide) blend*, Polym Deg Stabil 35:253-256.

[50] Azuma, Y., N. Yoshie, M. Sakurai, Y. Inoue, and R. Chujo (1992). *Thermal behaviour and miscibility of poly(3-hydroxybutyrate)/poly(vinyl alcohol) blends*, Polymer 33:4763 - 4767.

[51] Blumm, E., and A. J. Owen (1995). *Miscibility, crystallization and melting of poly(3-hydroxybutyrate)/poly(L-lactide) blends*, Polymer 36:4077-4081.

[52] Koyama, N., and Y. Doi (1997). *Miscibility of binary blends of poly[(R)-3-hydroxybutyric acid] and poly[(S)-lactic acid]*, Polymer 38:1589-1593.

[53] Koyama, N., and Y. Doi (1996). *Miscibility, thermal properties, and enzymatic degradability of binary blends of poly[(R)-3-hydroxybutyric acid] with poly[e-caprolactone-co-lactide]*, Macromolecules 29:5843-5851.

[54] He, Y., T. Masuda, A. Cao, N. Yoshie, and Y. Doi (1999). *Thermal, crystallization and biodegradation behavior of poly(3-hydroxybutyrate) blends with poly(butylene succinateco-butylene adipate) and poly (butylene succinate-co-e-caprolactone)*, Polym J 31:184- 192.

[55] Kumagai, Y., and Y. Doi (1992). *Enzymatic degradation of binary blends of microbial poly(3-hydroxybutyrate) with enzymatically active polymers*, Polym Deg Stabil 37:253-256.

[56] Kumagai, Y., and Y. Doi (1992). *Enzymatic degradation and morphologies of binary blends of microbial poly(3-hydroxybutyrate) with poly(e-caprolactone), poly(1, 4-butylene adipate) and poly (vinyl acetate)*, Polym Deg Stabil 36:241-248.

[57] Peoples, O. P., and A. J. Sinskey (1996). *Polyhydroxybutyrate polymerase*, U.S. Patent 5,534,432.

[58] Skraly, F. A., and O. P. Peoples (2001). *Polyhydroxyalkanoate biopolymer compositions*, U.S. Patent 6,323,010.

D. STARCHES, PROTEINS, PLANT OILS, AND CELLULOSICS

1. Introduction

Polysacharides, plant proteins, plant oils and lignin are four of the most widely available naturally occurring biomaterials, and they have therefore been the earliest materials utilized in bioplastics production. In fact, in their raw forms these plant-based materials have been exploited by humankind for millennia. Presently, starch is still the major component of approximately 75 percent of green plastics production if all biodegradable plastics are considered. Soy proteins have potential as adhesives and tremendous progress has been made in the past decade in using chemical methods to convert soy and other plant oils into useful materials. Cellulose, in turn, is emerging as a valuable component of bioplastics as a reinforcing member of composites, lending strength, durability, and heat tolerance to various biobased polymers. A brief discussion of these bioresources is presented here for completeness, however, for many of the materials discussed in this section, chemical manipulation is the predominant technology used to produce useful plastic materials. Accordingly, the emphasis is placed on the more promising areas of investigation involving industrial bioengineering techniques. For many of the materials within this broad class, the predominant bioengineering occurs in the genetic manipulation of the plant to produce a more desirable starting feedstock. DuPont's high oleic soybean is a good example of this type of manipulation; the soybean has been genetically engineered to produce a more desirable distribution of soy oils–in this case, one particularly rich in oleic oil.

2. State of the Science

2.1. Thermoplastic Starch (TPS)

Starch, an energy storage material found abundantly in cereals and tubers, was one of the first natural polymers studied for the production of biodegradable materials and is widely used in the food, paper, and textile industries. It is composed of amylose and the branched polymer amylopectin, both are polysaccharides of alpha-D-glucopyranosyl units linked by (1-4) and (1-6) linkages [1].

Starch was first used in biodegradable materials in the 1970s as a filler in synthetic polymers such as polyethylene or, in its gelatinized form, as a component of blends with water- soluble or water-dispersible polymers. These were best described as bio-disintegratable rather than biodegradable; data showed that only the surface starch was decomposed, leaving behind recalcitrant polyethylene fragments. Because products made from these resins did not meet the criteria of biodegradability for defined disposal systems like composting, further applications were soon sought [2].

More recently, starch has been used as the primary component in thermoplastic compositions. Although starch is not itself a thermoplastic material, at moderately high temperatures (90–180°C), under pressure and shear stress, starch granules melt and flow to give an amorphous material called TPS, which can be processed just like a thermoplastic synthetic polymer. Conventional plastic processing techniques such as injection molding and extrusion can then be applied successfully. The ability to mold TPS, and the advantages of starch in terms of low cost and high availability from renewable resources, have made it a highly attractive resource for the development of biodegradable polymers. Unfortunately, however, the use of regular TPS has been limited by its brittleness, by degradation under conditions of normal use, and by hydrophilicity [1, 2)] The latter is especially problematic because water is a plasticizer of starch, with the result that the performance of regular TPS is unstable at sufficiently great relative humidity [3].

To address these problems, a number of modifications of TPS have been attempted to improve its material properties. Surface modifications form one class of approaches; in these, the superficial hydroxyl groups of the material are derivatized, especially to reduce hydrophilicity, without changing the bulk composition and characteristics of the TPS. This type of approach has also been used with cellulose fibers, considered below, to improve their compatibilities with polymer matrices when they are used as natural reinforcements in composites [1].

The use of biodegradable plasticizers has also been investigated, with the goals of increasing durability as well as diminishing brittleness. Polyols such as glycerol, which is often used with biodegradable polymers, effectively reduce degradation of thermoplastic starch with increasing glycerol content corresponding to diminishing starch degradation under conditions of normal use [4].

The starch structure can also be modified directly, particularly to increase durability and plasticity. Among the many possibilities of modification, esterification is one of the most important ones. Direct esterification is the simplest route, but the unavoidable degradation of starch chains diminishes mechanical strength of the end product. Alternatively, starch can be covalently linked to other materials during a polymerization process; starch-poly(vinyl acetate) materials, for example, have been prepared via in situ polymerization of vinyl acetate in the presence of starch with a ferrous ammonium sulfate-hydrogen peroxide redox initiator system. Other methods include melt blending of starch with synthetic polymers, such as poly(ethyleneco-vinyl acetate) and polyethylene with anhydride functionality [3].

Current work in this area is directed toward the modification of TPS by reactive blending with polymers containing functional groups, which are intended to bond with starch hydroxyl groups during the blending procedure by means of a polymer analog (trans-esterification) reaction. Poly(vinyl acetate) and poly(vinyl acetate-co-butyl acrylate) are of special interest because of their potentials to diminish moisture sensitivity and the glass transition temperature of resulting blends. Preliminary experiments have shown successful reactive blending, increases in thermal stability, and decreases in swelling due to water, although work remains to be done to improve the processability of the blends [3].

In addition, mathematical and computational efforts involving models such as the lattice-fluid hydrogen-bonding model are assisting experimental work by facilitating the prediction of mechanical and volumetric properties of starch-based polymers and water [3].

Transgenic plants again offer the hope of improving polymer properties. Transgenic plants have been studied to understand the biosynthesis of starch [5, 6]. Manipulation of these biosynthetic pathways provides a means for affecting the distribution of amylose and amylopectin, potentially providing exquisite control over material properties within a single crop without the need to blend different varietals.

Several other polysaccharides are valuable bioresources from a materials point of view [7]. Naturally this includes cellulose, but also includes pectins. Pectins are used in the food industries as coatings and additives and can be extracted from apple pomace and citrus peels; pectin is partially methylated poly-á-1,4-D-galacturonic acid. Konjac (a copolymer of mannose and glucose with a ratio of about 1.6:1) is derived from plant tubers and can be formed into films. Alignates are also film formers that are derived from the cell wall of brown seaweed; structurally alignate is poly(1,4 uronic acid). Guar gum and gum Arabic are two widely recognized materials from the family of plant gums that consist of hydrateable polysaccharides; these gums find application as binders, adhesives, flocculants, emulsifiers, and even lubricants in the food, papermaking, and petroleum industries.

Polysacharides are also available from animal sources; the primary example is chitin. Chitin is the second only to cellulose in its natural abundance in biomass being found in the exoskeletons of crustacean and insects. Because chitin and the related chitosan are available from shellfish waste, they are inexpensive in their raw forms and have been widely studied [8]. Applications include absorbants used in wastewater treatment, films used as membranes, beads for metal chelation, and coatings for improved seed germination. To date, most manipulation of the raw materials to final products has involved chemical methods with relatively little emphasis on bioengineering techniques. Given the abundance of chitin, better more benign processing techniques are warranted.

From the perspective of bioengineering, bacterial polysaccharides are the most interesting as they lend themselves to the host of metabolic engineering techniques. For example, xanthan gum is a high molecular weight branched polysaccharide extracted from Xanthomonas campestris that is used as a viscosity modifier in drilling fluids. By introducing mutants in various parts of the biosynthetic pathway through genetic engineering, a series of polymers with different rheological properties can be produced by fermentation [9, 10].

2.2. Plant Proteins

Several potentially useful plant proteins are available as byproducts of biorefining. For example, corn zein (the alcohol soluble fraction of corn gluten) is becoming increasingly available as a result of expansion in the production of bioethanols. Protein based adhesives appear promising because many proteins in nature are used for adhesion, for example, marine mussels use proteins to adhere to surfaces.

Widespread cultivation of soybeans in the United States has encouraged a great deal of research into the development of biopolymers derived from soy protein. Soy proteins are complex macromolecules with many sites available for interaction with plasticizers and other copolymeric constituents, enabling soy protein to be converted to soy protein plastic through extrusion with a plasticizer or cross-linking agent. Although the mechanical properties of soy protein plastic can be controlled and optimized by adjusting the molding temperature and pressure and initial moisture content, applications are limited because of its low strength and great tendency to absorb moisture. The most effective method is to blend soy protein plastic with biodegradable polymers to form soy protein-based biodegradable plastics. Currently, the biodegradable plastics being used to blend soy plastic include polyester amide, polycaprolactone, Biomax®, and poly(tetramethylene adipate-co-terephthalate) [11, 12]. While these approaches have been effective, other promising options are found in the use of soy proteins in structural composites, addressed below.

Proteins can be synthesized using recombinant DNA technologies. Because of this there is great interest within the polymer science community in exploiting proteins to make specialized supramolecular structures. The primary protein sequence of amino acids (residues) dictates its molecular conformation and resulting supramolecular structure. If this structure formation can be controlled novel and interesting materials will result. It is unclear however, if such materials will have widespread utility as commodity plastics.

2.3. Plant Oil-based Polymers

Soy bean oil and other plant oils have already proven useful as plastic materials when converted using chemical techniques. Their great utility lies in the fact that the much of the oil present in plants is unsaturated (i.e., the oils contain reactive carbon double bond) and contain ester linkages. These chemical functionalities allow a range of polymerization chemistries to be practiced leading to a wide variety of plastic resins. From a bioengineering point of view, genetic engineering techniques may be used to manipulate and control the distribution of the type of oil present in the plant. DuPont has developed a soybean that contains in excess of 80 percent oleic within the fatty acid distribution. In 2002, 75 percent of U.S.-soybean acres were planted with biotech soybeans–up from 68 percent in 2001, according to statistics released by the USDA [13]. The widespread acceptance of transgenic crops likely means that plant-based oils of defined character will become increasingly available in the future. Once again, the conversion of these specialty plant oils to plastics is expected to be more economical than conversion to biodiesel.

Natural oils are comprised of triglycerides and are abundant in many areas of the world. Triglycerides consist of three fatty acids (i.e., carbon numbers from 22 down to 14 and double bonds down from 3 to 0) covalently linked to a central glycerol.

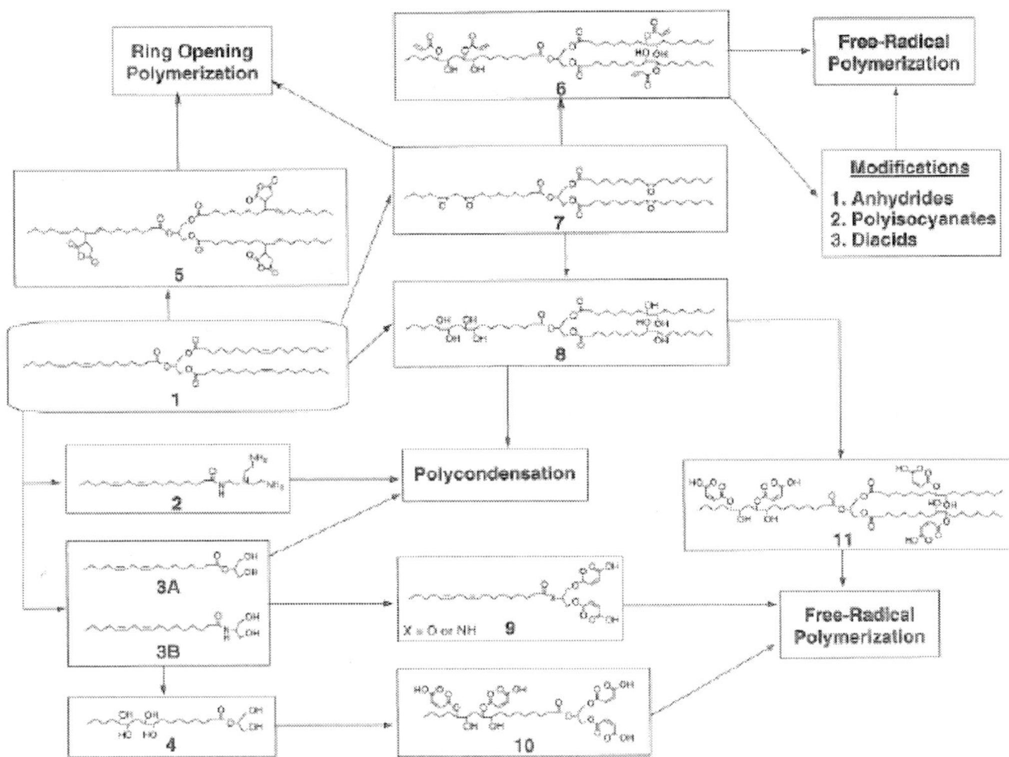

Figure 13. Pathways for the chemical conversion of plant triglycerides (1) into different classes of chemically reactive species useful for making plastics (from Wool [14]).

A variety of chemistries is available for turning such oils into useful polymeric products and these are discussed at length in the recent compilation by Wool from which Figure 13 is taken [14]. The various functionalities available include amines (structure 2), monoglycerides (3A, 3B), polyols [4], maleates [5], acrylates [6], epoxy [7], hydroxyl (8), and maleate half esters [9, 10, 11]. Such monomers are easily incorporated into well established vinyl ester resin formulations and can serve to decrease volatile organic compound (VOC) emissions in styrene containing formulations. Other polymeric materials that can contain significant plant-oil content include polyurethanes

2.4. Cellulose, Lignin, and Reinforced Composites

As is evident from the preceding sections, many plastics can be improved substantially for materials applications by the introduction of reinforcing constituents to form composites. Indeed, certain biosubstances such as proteins can only form usable materials as components of composites. In the interest of minimizing environmental impact, natural cellulose-based fiber reinforcements have become especially attractive.

2.4.1. Cellulose. Cellulose, the most abundant biological substance on Earth, is a polymer of glucose with both amorphous and crystalline regions in its native state (see Chapter IV. Bioethanol for additional structural details). Although cellulose is most widely applied as a reinforcing agent in composites, it can also be used directly or as an esterified derivative in place of petroleum as a feedstock to make cellulosic plastics, particularly as flexible films. The most widely used cellulose esters are cellulose acetate (CA), cellulose acetate propionate

(CAP), and cellulose acetate butyrate (CAB), [15] although cellulose fatty esters are under investigation as well [16]. These later esters can be produced from wastes such as recycled paper and bagasse, and have found applications in film substrates for photography, toothbrush handles, selective filtration, adhesive tapes, cellophane, various semi-permeable and sealable films, and automotive coatings [17] (http://www.innoviafilms.com/product/summary _datasheets.htm).

2.4.2. Lignin. The pulp and paper industries produce in the vicinity of 20 million tons per year of lignin in the United States alone [18]. Along with cellulose and hemicellulose, lignin constitutes a major component of the cell wall in both hardwoods and softwoods; lignin content ranges from 15–40 percent by weight [19]. Lignins are complex heteropolymers of p-hydroxy cinnamyl alcohols. Lignin is derived by treatment of pulp with sodium hydroxide and sodium sulfide (kraft pulping liquor) which cleaves linkages in the protolignin structure [20]. Glasser and Lora have recently published a survey of lignin useages [21]. Lignin can substitute for petroleum based materials in a number of ways and an excellent overview of these uses is available [14]. Lignin can serve as a filler in thermoplastics, thermosets, and rubbers; it has also been converted into carbon fibers, a potentially high value-added application. While most efforts in lignin modification have relied on chemical methods, enzymatic grafting has been reported [22]. In the context of the integrated biorefinery, if forest resources (hardwoods and softwoods) are to be exploited in the future, additional uses for lignin will be needed. Presently it is typically burned as a fuel but higher value added uses are clearly desirable but technically challenging.

For example, downregulation of lignin content is of considerable interest. Transgenic aspen trees in which expression of a lignin biosynthetic pathway has been downregulated by antisense inhibition have been developed [23]. However, while these trees produced up to 45 percent less lignin, this was accompanied by a 15 percent increase in cellulose. That is, the lignin to cellulose mass ratio remained essentially unchanged.

2.4.3. Natural fiber-reinforced composites. Natural fibers such as kenaf, flax, jute, hemp, sisal, and henequen are attractive options for reinforcing starch- and protein-based composites because they are renewable and sustainable, as well as low cost, with acceptable mechanical properties, ease of separation, and biodegradability. Additionally, these fibers have excellent thermal and sonic insulation properties. Natural fibers from grass, hemp, and ramie have been reported as reinforcements for soy-based matrices, compared with glass fibers, and improved in terms of physical properties via surface treatments [11, 12, 24, 25].

Although both soy plastics and natural fibers possess hydroxyl and carboxyl groups that can interact during processing, these interactions seem to be limited and typically do not lead to significant improvements in performance. However, enhancing these interactions through the use of compatibilizers, molecules such as polyester amide (PEA) that interact with both the fiber and the polymer matrix, is of current interest in the synthesis of natural fiber-reinforced soy biocomposites. Among reinforcing fibers, pineapple leaf fiber is an attractive option due to its high tensile strength (400–1600 MPa) and modulus (59 GPa), that result in turn from its high cellulose content (70–82 percent) and high degree of crystallinity. In addition, its cultivability in the southern United States and Central and South America cause it to be widely available, and it has shown previous success in strengthening low-density polyethelyne (LDPE), rubber, polyester, and polyhydroxybutyrate (PHB) [11, 12].

Hemp fibers and related grass fibers are additional promising natural reinforcement agents. Hemp is an important ligno-cellulosic natural fiber that contains (by weight) 70.2–

70.4 percent cellulose, 3.7–5.7 percent lignin, 17.9–22.4 percent hemicellulose, 0.9 percent pectin, 0.3 percent wax, and 10.8 percent moisture. Natural waxy substances on the fiber surfaces diminish fiber-matrix bonding, leading to the necessity of investigation of various surface treatments to improve fiber-matrix adhesion of the resulting biocomposites. In addition, hemp, as a natural fiber, starts to decompose at 3 00°C, with the consequence that its use is limited to plastics that can be processed at lower temperatures. Fortunately, that is not an obstacle for soy- and starch-based plastics [17].

2.4.4. Nanocomposites. Particles are often incorporated into plastics to improve stiffness and toughness, to enhance barrier properties or fire resistance, or simply to reduce cost. Unfortunately, however, brittleness and opacity are sometimes imparted to the resulting composites. Nanocomposites are a new class of plastics that incorporate nanoparticles, or particles having at least one dimension in the nanometer range, in attempts to reduce these undesirable side effects.

2.4.4.1. Classes. Three types of nanocomposites are distinguished based on the number of dimensions of the dispersed particles that are in the nanometer range. If all three dimensions are in the nanometer range, the nanoparticles are isodimensional; examples include spherical silica [26] and other [27] nanoparticles. Nanotubes or nanowhiskers are particles in which only two dimensions are in the nanometer scale; examples include carbon nanotubes [28] and cellulose whiskers [29–31]. If only one dimension is on the nanometer scale, the filler is present in the form of sheets of a few nanometers thick. These are polymer-layered crystal nanocomposites and are almost exclusively obtained by the intercalation of the polymer inside galleries of layered host crystals. A wide variety of both synthetic and natural crystalline fillers are able to intercalate as polymers. Those based on clays and other layered silicates are most widely investigated because the starting materials are widely available and inexpensive. In addition, the intercalation chemistry of clays has been well-studied [32, 33]. When such sheet- shaped nanofillers are successfully dispersed in a polymer, the resulting nanocomposites exhibit markedly improved mechanical, thermal, optical, and physio-chemical properties when compared with the pure polymer or conventional (microscale) composites. For example, the first profound demonstration provided by Kojima and coworkers [34–37] for Nylon6-clay nanocomposites showed that improvements can include increased moduli, strength, and heat resistance, as well as decreased gas permeability and inflammability.

2.4.4.2. Cellulose nanocomposites. In recent years, natural cellulose fibers have gained attention as reinforcing phases for polymer nanocomposites [30, 31, 38–46]. Several approaches to the production of such microfibers are known, including chemical treatments and steam explosion of cellulose starting materials. Their low density, gentleness toward processing equipment, and relatively reactive surfaces hold great promise for excellent property improvement. In addition, they are abundant and inexpensive, and most notably, they retain their biodegradability [47].

In cellulose nanocomposites, the polymer modulus can be increased by more than three-fold at 6 percent loading levels due to the long aspect ratio of the cellulosic fillers. These fibers reach the percolation threshold at relatively low loading, causing the modulus to increase rapidly. Cellulose nanocrystals can exhibit high aspect ratios, whereby the length divided by thickness can approach 500 for some agricultural fibers such as sugar beet and the giant reed Arundo donax. Proper treatment can also lead to the desired platelet-like morphology similar to the very successful clay nanofillers [48].

In efforts to synthesize completely biodegradable nanocomposites, so-called green nanocomposites have been successfully fabricated from cellulose acetate powder, biodegradable triethyl citrate (TEC) plasticizer, and organically modified clay. Varieties with 20 weight percent plasticizer and 5 weight percent organoclay showed better intercalation and an exfoliated structure than the counterpart having 30–40 weight percent plasticizers, while the tensile strength, modulus, and thermal stability diminished with increasing plasticizer content from 20-40 weight percent. Of special note, the nano-reinforcement at the lower volume fractions (phi ≤ 0.02) reduced the water vapor permeability by 50 percent [49].

2.4.4.3. Polysaccharide nanocomposites. Renewable nanocomposites have also been made using microbially-derived polysaccharide polymers, such as polysaccharide fillers as the reinforcing phase for a poly(3HO) [45, 50]. The mechanical properties of poly(3HO) are significantly improved by the addition of certain polysaccharides (up to 50 percent starch by weight or up to 6 percent cellulose by weight). The aspect ratio of the reinforcing phase is a critically important variable. Improved nanocomposites are possible by transreacting the hydroxyl groups of polysaccharides with functional groups in PHA, again demonstrating the importance of chain functionalization [45, 50].

3. Research Priorities

3.1. Basic Biosciences

From a long-term perspective, continued support for basic biosciences that allow the manipulation of plants at the genetic level is absolutely essential. Only through continued development of genetic and physiological engineering techniques will the possibility of inducing organisms to produce polymers directly, and thereby reducing the cost of bioplastics by minimizing processing steps, be realized. In addition, the preference for anaerobic processes among microbial fermentations should be recognized due to the minimization of carbon loss from feedstocks. In the context of the plant-based plastics described in this section, the required tools include the genetic engineering of the cellulose-lignin and oil distribution in plants. Such basic genetic manipulation of plants is supported by the USDA. Specific strategies for pollution prevention are necessarily more narrow and should focus on developing cost-effective methods for producing plastics from plant-based matter.

3.2. Biodegradable Plastics

Within the realm of starch-based plastics, commercial interest and success is currently carrying the development of compostable disposable packaging, including garbage bags, food wraps, diapers, as well as disposable food service items such as plates, cups, and utensils. Commercial research and development is even leading to improvements in water resistance and durability of starch-based materials, with the result that these are not considered to be high priorities for federal research funding. In contrast, the possibility of vastly improved strength, lightness, durability, and heat and water resistance offered by natural fiber-reinforced composites, particularly nanocomposites, cause this area to be highly attractive for additional effort. Low-cost polysaccharide-based plastic materials have the greatest potential to displace significant amounts of petroleum-based plastics.

Similarly, more benign greener chemical processes should be investigated for transforming the available renewable resources. Such activities should include enzymatic transformations when they are less energy intensive than existing chemical routes.

Genetic engineering of plant proteins for specific functionality is a lower priority simply because the potential application, adhesives, is small relative to the packaging and structural uses for plastics. If plant proteins can be engineered for larger volume applications, for example, into thermoplastic film or sheet materials, the promise of coproducing both fuels and materials in an integrated biorefinery would be supported. Specialized materials produced through recombinant DNA technologies are expected to be of high commercial value but of relatively small volume.

4. Commercialization

Commercialization of many of the bioplastics discussed in this section is well established. Chemically modified cellulosic esters are used in a wide range of products. Importantly, starch- and cellulose-based products are meeting increasingly high standards of biodegradability and compostability, as prescribed by the American Society for Testing and Materials (ASTM), DIN CERTCO in Germany, and EK certification in Norway. Representative of these, ASTM D6400-99 "Specifications for Compostable Plastics" or ASTM D6868 "Specification for Biodegradable Plastic Coatings on Paper and other Compostable Substrates" certification designates "a plastic or substrate that undergoes degradation by biological processes during composting to yield CO_2, water, inorganic compounds, and biomass at a rate consistent with other known compostable materials and leaves no visible, distinguishable or toxic residue" (www.astm.org). Commercial starch- and cellulose- based materials are highly visible worldwide, and their producers and distributors, including BASF, Novamont, PolarGruppen, Bio-Bag Canada, Innovia Films, and Biosphere Products Inc., to name a few, are prominent members of the Biodegradable Products Institute (www.BPIWorld.org).

4.1. Mater-Bi™

Mater-BiTM is one of the most prominent starch-based plastics, manufactured by Novamont Corporation in Terni, Italy. This product is composed primarily of cornstarch, complexed with proprietary natural and/or synthetic polymers to impart water-resistance and durability, and is produced in pellets of three grades that can be processed by distributors using common transformation techniques. For recommended uses, including packaging, disposable plastics, toys, and biofillers, the company claims that Mater-Bi™ plastics have "characteristics [that] are identical to those of traditional plastics, but are perfectly biodegradable and compostable" (www.materbi.com). After use, products made of Mater-Bi™ biodegrade on average, in the time of one composting cycle. Although Mater-Bi™ was developed only in the late 1980s, and Novamont was founded only in 1990, Mater-Bi™ products have already become internationally visible. The product called Green Pen, made of Mater-Bi™ by the company Lecce Pen (Turin, Italy), was chosen to be the official pen of the United Nations Earth Summit in Rio De Janeiro in 1992, and in September of the same year, the production of the first MaterBiT^M bags for separate waste collection began in Germany. Research collaborations with Goodyear Tire began in 1995, and in 2001 production of the

GT3 tire began, using Biotred technology that uses biofillers made of Mater-Bi^{TM}. Production of Mater-Bi^{TM} diapers, by the Swedish company Naty, began in 1999, and in 2000, the Olympics in Sydney, Australia used catering products and compost bags made of Mater-Bi^{TM} (www.materbi.com).

Mater-Bi^{TM} is processed and distributed by numerous companies worldwide, including, in addition to those named above, NAT-UR in California, which also distributes NatureWorks PLA-based materials (www.nat-ur.com), as well as PolarGruppen and Bio-Bag Canada, which manufactures biodegradable, compostable bags and packaging from Mater-Bi^{TM} (www.polargruppen.com; www.biobag.ca) that are largely ASTM, DIN CERTCO, and EK certified.

4.2. Ecoflex®

In 1998, the German company BASF introduced a family of biodegradable aliphatic-aromatic copolyesters, including formulations consisting of thermoplastic starch, under the brand name Ecoflex®. Ecoflex® is recommended for trash bags and disposable packaging, as well as for starch composites, since it degrades completely in compost within just a few weeks. Homopolymeric Ecoflex®, lacking the starch component, has been developed specifically for flexible film applications, such as those commonly filled by polyethylene, and is now established as one of the first completely biodegradable flexible films. Like polyethylene, Ecoflex® is waterproof, tear resistant, flexible, fusible, and imprintable, and appears to have high processing flexibility (http://www2.basf.de/basf2/html/plastics/englisch/pages/biokstoff/ecoflex.htm). In 2005, BASF commercialized blends of Ecoflex® with PLA under the tradename Ecovio®.

4.3 PPM

PPM was developed by Biosphere Industries from cellulose-reinforced starch to meet rigid packaging needs and has two primary grades, PPM 100 and PPM200, which are currently used primarily in food-service items and general packaging. PPM 100 biodegrades 98 percent in 28 days (100 percent in less than 40 days) and can resist warm-water extended periods. For greater heat and moisture resistance, the higher-grade PPM200 is used; this material requires more than 40 days for complete compostability but safely withstands hot liquids for extended periods. In January 2005, PPM 100 received Biodegradable Products Institute certification, demonstrating that the material meets the ASTM D6868 specifications and will biodegrade swiftly and safely during municipal, commercial, or household composting (www.biospherecorp.com).

4.4. Cellulose Films

Innovia Films, Inc. was formed in 2004 and is one of the producers of cellulose-based packaging films. The variety of products offered includes nitrocellulose-coated films, laminating films, films for adhesive tape, photographic masking, and heat-sealable films for microwave and oven applications (http://www.innoviafilms.com/product/summary _datasheets.htm).

4.5. Soy-based Plastics

The Soy Works Corporation in Illinois holds a number of patents for soy-protein-based products and has developed a soy-protein-based plastic known as SoyPlusTM. Resins are made of soy protein and other unidentified ingredients, yielding a moldable resin that is being developed for a variety of applications, including garden supplies, food-service items, industrial packaging, mulch, toys, golf tees, and building materials (http://soy workscorporation.com/).

Another soybean product, based from oil rather than protein, was developed by the University of Delaware and is a primary ingredient in molded fiberglass-reinforced farm equipment parts being tested by the John Deere Company (http://www.unitedsoybean.org/feedstocks/fsv2i6b.html). This work has continued and resulted in the founding of Cara Plastics (http://www.caraplastics.com). Cara Plastics is pursuing market opportunities in structural composites, primarily with polyesters and vinyl esters where Cara's resins would be used as the matrix of a fiber-reinforced material. Markets are estimated to be 2 billion pounds worldwide compared to 500 million pounds for biodegradable plastics. Cara hopes to ultimately capture at least 10 percent of these markets, or 250 million pounds annually with a target of 75 million pounds annually within 5 years.

REFERENCES

[1]   Carvalho, A. J., A. A. Curvelo, and A. Gandini (2005). *Surface chemical modification of thermoplastic starch: reactions with isocyanates, epoxy functions and stearoyl chloride,* Indust. Crop Product 21:33 1-336.

[2]   Stevens, E. S. (2005). Green plastics: Environment-friendly technology for a sustainable future, http://www.greenplastics.com/.

[3]   Vargha, V., and P. Truter (2005). *Biodegradable polymers by reactive blending trans-esterification of thermoplastic starch with poly (vinyl acetate) and poly (vinyl acetate-cobutyl acrylate),* Eur Polym J 41:715-726.

[4]   Rahman, M., and C. S. Brazel (2004). *The plasticizer market: An assessment of traditional plasticizers and research trends to meet new challenges,* Progr Polym Sci 29: 1223-1248.

[5]   Ball, S., H.-P. Guan, M. James, A. Myers, P. Keeling, G. Mouille, A. Buleon, P. Colonna, and J. Preiss (1996). *From glycogen to amylopectin: a model for the biogenesis of the plant starch granule,* Cell 86(3):349-353.

[6]   Stark, D. M., K. P. Timmerman, G. F. Barry, J. Preiss, and G. M. Kishore (1992). *Regulation of the amount of starch in plant tissues by ADP glucose pyrophosphorylase (adenosine diphosphate),* Science 258(5080) :287-293.

[7]   Kirk- *Othmer Encyclopedia of Chemical Technology,* 5th Edition (2005). John Wiley & Sons, Inc.

[8]   Hudson, S. M., and C. Smith (1998). *Polysacharides: Chitin and Chitosan: Chemistry and Technology of Their Use As Structural Materials, in Biopolymers from Renewable Resources* (Kaplan, D. L., Ed.). Springer, Berlin.

[9]   Hassler, R. A., and D. H. Doherty (1990). *Genetic engineering of polysaccharide structure: Production of variants of Xanthan Gum in Xanthamonas campestris,* Biotechnol Progress 6:182-187.

[10]   Yalpani, M. (1987). *Industrial Polysaccharides: Genetic Engineering Structure / Property Relations and Applications,* Elsevier, Amsterdam.

[11]   Liu, W., M. Misra, P. Askeland, L. T. Drzal, and A. K. Mohanty (2005). *'Green' composites from soy based plastic and pineapple leaf fiber: fabrication and properties evaluation,* Polymer 46:2710-2721.

[12]   Liu, W., A. K. Mohanty, P. Askeland, L. T. Drzal, and M. Misra (2004). *Influence of fiber surface treatment on properties of Indian grass fiber reinforced soy protein based biocomposites,* Polymer 45:7589-7596.

[13]   United States Department of Agriculture, National Agricultural Statistics Service (2002). usda.mannlib.cornell.edu/reports/nassr/field/pcp-bba/acrg0602.txt, June 28.

[14]   Wool, R. P., and X. S. Sun (2005). Bio-Based Polymers and Composites. Elsevier Academic Press, Burlington, MA.

[15]   Yoshioka, M., and N. Shiraishi (2001). *Biodegradable Plastics from Lignocellulosics, in Wood and Cellulosic Chemistry* (David Hon, N. S., Ed.). Marcel Dekker, New York, NY.

[16]   Satgé, C., R., B. Granet, P. Verneuil, Branland, and P. Krausz (2004). *Synthesis and properties of biodegradable plastic films obtained by microwave-assisted cellulose acylation in homogeneous phase,* Comptes Rendus Chimie 7:135-142.

[17]   Mohanty, A. K., A. Wibowo, M. Misra, and L. T. Drzal (2004). *Effect of process engineering on the performance of natural fiber reinforced cellulose acetate biocomposites,* Compos Part A: Appl Sci Manuf 35:363-370.

[18]   Chum, H. (1989). *Assessment of Biobased Materials.* Solar Energy Research Institute (SERI, now NREL) - SERI R-234-3610.

[19]   Fujita, M., and H. Harada (2001). *Ultrastructure and Formation of Wood Cell Wall, in Wood and Cellulosic Chemistry* (David Hon, N. S., Ed.). Marcel Dekker, New York, NY.

[20]   Sakakibara, A., and Y. Sano (2001). *Chemistry of Lignin, in Wood and Cellulosic Chemistry* (David Hon, N. S., Ed.). Marcel Dekker, New York, NY.

[21]   Lora, J. H., and W. G. Glasser (2002). *Recent industrial application of lignin: A sustainable alternative to nonrenewable materials,* J Polym Environ 10:39-48.

[22]   Mai, C., O. Milstein, and A. Hüttermann (2000). *Chemoenzymatical grafting of acrylamide onto lignin,* Journal of Biotechnology 79:173-183.

[23]   Hu, W.-J., S. A. Harding, J. Lung, J. L. Popko, J. Ralph, D. D. Stokke, C.-J. Tsai, and V. L. Chiang (1999). *Repression of lignin biosynthesis promotes cellulose accumulation and growth in transgenic trees,* Nature Biotechnology 17:808-812.

[24]   Joshi, S. V., L. T. Drzal, A. K. Mohanty, and S. Arora (2004). *Are natural fiber composites environmentally superior to glass fiber reinforced composites?* Composites Part A: Applied Science and Manufacturing 35:371-376.

[25]   Mishra, S., A. K. Mohanty, L. T. Drzal, M. Misra, and G. Hinrichsen (2004). *A review on pineapple leaf fibers, sisal fibers and their biocomposites.* Macromolecular Materials and Engineering 289:955-974.

[26] von Werne, T., and T. E. Patten (1999). *Preparation of structurally well defined polymer-nanoparticle hybrids with controlled/living radical polymerization,* J Am Chem Soc 121:7409-7410.

[27] Reynaud, E., C. Gauthier, and J. Perez (1999). *Nanophases in polymers,* Rev Metall/Cah Inf Tech 96:169-176.

[28] Calvert, P. (1997). *Potential applications of nanotubes, in Carbon Nanotubes* (T.W. Ebbesen Ed.). CRC Press, Boca Raton, FL, 277-292.

[29] Chazeau, L., J. Y. Cavaille, G. Canova, R. Dendievel, and B. Boutherin (1999). *Viscoelastic properties of plasticized PVC reinforced with cellulose whiskers,* J Appl Polym Sci 71:1797-1808.

[30] Favier, V., G. Canova, S. Shrivastava, and J. Cavaille (1997). *Mechanical percolation in cellulose whisker nanocomposites,* Polymer Engineering and Science 37(10):1732-1739.

[31] Favier, V., H. Chanzy, and J. Y. Cavaille (1995). *Polymer nanocomposites reinforced by cellulose whiskers,* Macromolecules 28(18): 6365-6367.

[32] Grim, R. E. (1953). *Clay Mineralogy.* Mc-Graw Hill, New York, p 384.

[33] Theng, B. K. G. (1974). *The Chemistry of Clay-Organic Reactions.* Wiley, New York.

[34] Kojima, Y., A. Usuki, M. Kawasumi, A. Okada, Y. Fukushima, T. Karauchi, and O. Kamigaito (1993). *Mechanical properties of nylon-6-clay hybrid,* J Mater Res 6:1185-1189.

[35] Kojima, Y., A. Usuki, M. Kawasumi, A. Okada, T. Kurauchi, and O. Kamigaito (1993). *Synthesis of nylon 6-clay hybrid by Montmorillonite Intercalated with e-Caprolactam,* J. Poly Sci A: Poly Chem 3 1:983-986.

[36] Kojima, Y., A. Usuki, M. Kawasumi, A. Okada, T. Kurauchi, and O. Kamigaito (1993). *One-pot synthesis of nylon 6-clay hybrid,* J Poly Sci A: Poly Chem 31:1755-1758.

[37] Kojima, Y., A. Usuki, M. Kawasumi, A. Okada, T. Kurauchi, O. Kamigaito, and K. Kaji (1994). *Fine structure of nylon-6-clay hybrid,* J Poly Sci B: Poly Phys 32:625-630.

[38] Angeles, M. N., and A. Dufresne (2000). *Plasticized starch/tunicin whiskers nanocomposite materials,* Macromolecules 33:8344.

[39] Angeles, M. N., and A. Dufresne (2001). *Plasticized starch/tunicin whiskers nanocomposite materials.* 2. Mechanical Properties, Macromolecules 34.

[40] Cavaille, J. Y., H. Chanzy, V. Favier, and B. Ernst (2000). *Cellulose microfibril-reinforced polymers and their applications,* U. S. Patent 6,103,790.

[41] Cavaille, J. Y., H. Chanzy, E. Fleury, and J. F. Sassi (2000). *Surface-modified cellulose micro fibrils, method for making the same, and use thereof as filler in composite materials,* U.S. Patent 6,117,545.

[42] Cavaille, J. Y., and A. Dufresne (1998). *Nanocomposite materials of thermoplastic polymers reinforced by polysaccharide fillers, in Biopolymers*: Utilizing Natures Advanced Materials (Greene, R. V., S. H. Imam Eds.). ACS Publishing, New York.

[43] Dufresne, A. (1998). *High performance nanocomposite materials from thermoplastic matrix and polysaccharide fillers,* Recent Research Developments in Macromolecules Research 3(2):455-474.

[44] Dufresne, A., J.-Y. Cavaille, and W. Helbert (1996). *New nanocomposite materials: microcrystalline starch reinforced thermoplastic,* Macromolecules 29:7624-7626.

[45] Dufresne, A., M. B. Kellerhals, and B. Witholt (1999). *Transcrystallization in mclPHA s/Cellulose whisker composites,* Macromolecules 32:7396-7401.

[46] Dufresne, A., and M. R. Vignon (1998). *Improvement of starch film performances using cellulose microfibrils,* Macromolecules 31:2693-2696.

[47] Frisoni, G., M. Baiardo, and M. Scandola (2001). *Natural cellulose fibers: Heterogeneous acetylation kinetics and biodegradation behavior,* Biomacromolecules 2:476-482.

[48] Grunert, M., and W. Winter (2000). *Progress in the development of cellulose reinforced nanocomposites,* Polymeric Materials, Science and Engineering 82:232-23 8.

[49] Park, H. M., M. Misra, L. T. Drzal, and A. K. Mohanty (2004). *"Green" nanocomposites from cellulose acetate bioplastic and clay: Effect of eco-friendly triethyl citrate plasticizer,* Biomacromolecules 5:2281-2288.

[50] Dubief, D., E. Samain, and A. Dufresene (1999). *Polysaccharide microcrystals reinforced amorphous poly(b-hydroxyoctanoate) nanocomposite materials,* Macromolecules 32:5765-5771.

Chapter IV

BIOFUELS

The worldwide depletion of fossil fuels and widespread concern over increasing atmospheric CO2 have sparked interest not only in biomaterials but also in sustainable, non-fossil-based fuels. Political instability in petroleum-producing regions has further increased the desirability of domestic fuel sources, particularly for transportation [1]. Solar and wind power are well-suited to sustainable generation of electricity, including electricity for charging vehicle batteries, but most modern vehicles are designed for liquid fuels that are best simulated by two biofuels: bioethanol and biodiesel [2]. In addition, biohydrogen is an emerging biofuel that carries energy from sunlight or organic matter, rather than petroleum, in clean-burning hydrogen (H_2), Finally, biodesulfurization of petroleum products may offer a way to mitigate some effects of petroleum use during a transition and is discussed as well.

A. BIOETHANOL

1. Introduction

Proponents of ethanol as a use for fuel highlight the apparent net-zero contribution of fuel ethanol combustion to the global carbon cycle, in that feedstocks for ethanol production derive their carbon from atmospheric CO_2, and that ethanol combustion simply returns the fixed carbon to its atmospheric source [3]. Others, however, make the valid counterpoints that the conversion even of agricultural wastes to ethanol is, itself, an energy-intensive process that frequently makes use of fossil energy sources, and that growth of crops dedicated to energy production must also be conducted in a sustainable manner for fuel ethanol use to carry a net environmental benefit [4].

Petroleum currently supplies 97 percent of the energy consumed for transportation [5], and transportation accounted for two-thirds of U.S. petroleum use in 2002. This trend is expected to continue until 2025 [6]. This need not continue, however, as all automobile manufacturers produce flexible-fuel vehicles (FFVs) that can use 10 percent or 85 percent ethanol blends with gasoline, and ethanol can also replace diesel fuel in heavy vehicles [5]. The United States also now has 199 fueling stations for ethanol, as well as extensive online services for planning travel between stations [7]. Although ethanol is limited in availability in some states, the transportation market for ethanol could expand to as much as 3 8–53 billion liters per year, if all available agricultural residues were converted to ethanol [8]. Ethanol is also being used as a replacement for methyl tertiary butyl ether (MTBE), the fuel oxygenate that is being phased out due to its widespread contamination of groundwater [9].

The majority of ethanol, approximately 62 percent of the world total, is currently produced in Brazil, primarily from cane sugar (~12.5 billion liters in 2002), and in the United States, primarily from corn (~5 billion liters per year) [5, 10]. However, these feedstocks are expensive and are useful as foods, causing a great deal of research to be focused on the development of biomass such as corn cobs and stalks, sugar cane waste, wheat and rice straw, other agronomic residues, forestry and paper mill discards, paper municipal waste, and dedicated energy crops into ethanol [11]. While the use of such non-food substrates helps the economics of ethanol production substantially, the high cost of production, especially relative to gasoline, remains the primary obstacle to bioethanol commercialization [5].

Lignocellulosic (non-food) raw materials, such as agricultural, wood chip, and paper wastes, can yield approximately 100 billion gallons of fuel-grade ethanol per year in the United States alone [12]. The projected cost of bioethanol has dropped from about $1.22 per liter to about $0.31 per liter based on consistent improvements in pretreatment, enzyme application, and fermentation [13]. If additional specific improvement targets are met, this cost could drop to as low as $0.20–$0.12 per liter by 2015 [14]. For transportation fuel, therefore, ethanol has real potential to replace gasoline, even in the absence of governmental support.

2. State of the Science

2.1. Ethanol Biosynthesis: Overview

2.1.1. Feedstocks. Biological production of ethanol first requires that atmospheric CO_2 be fixed into organic carbon (biomass) through photosynthesis. While agricultural crops and residues currently form the vast majority of feedstock for ethanol production, other biomass sources such as wood chips, sawdust, industrial organic wastes, and municipal organic wastes are important for the commercial development of fuel bioethanol [11]. Agricultural plant matter contains approximately 10–15 percent lignin, a polymer of phenolic subunits that is highly resistant (although far from impervious) to enzymatic attack. Lignin typically surrounds and protects the more enzymatically-vulnerable components of cellulose and hemicellulose, which comprise approx. 20–50 percent and 20–3 5 percent of the remaining plant material, respectively [9].

The next challenge in bioethanol synthesis is therefore the release of fermentable sugars from the biomass, or conversion of the feedstock into fermentable substrates. This phase involves both the separation of lignin from cellulosic and hemicellulosic polymers and the hydrolysis of the polymers into monomeric sugars, primarily glucose and xylose. This phase is termed pretreatment, and a variety of biotic and abiotic approaches are currently under investigation; abiotic approaches have been recently reviewed [15].

2.1.2. Mechanical and chemical disruption. Lignin is typically dissociated from the carbohydrates by mechanical and/or thermochemical means, including hot water, steam explosion, and/or acid treatments in either batch or flow-through reactors. Many variations have been explored. While it was once difficult to compare the performance and economics of the various approaches due to differences in feedstocks tested, a group of pretreatment researchers has formed in North America to facilitate such comparisons. This group, the

Biomass Refining Consortium for Applied Fundamentals and Innovation (CAFI), has the goal of advancing the efficacy and knowledge base of pretreatment technologies (3), reviewed in [16, 17].

Mechanical lignin disruption effectively hydrolyzes a significant fraction of the hemicellulose, but is less effective in hydrolyzing cellulose. This difference is caused by the different structures of the two polymers: hemicellulose is a highly branched, typically amorphous polymer that is therefore relatively easy to hydrolyze into its component sugars (pentoses D-xylose and L-arabinose; hexoses D-galactose, D-glucose, D-mannose; and uronic acid; all highly substituted with acetic acid). Hemicelluloses from hardwoods are typically high in xylose, while those in softwoods contain more hexoses [3].

Cellulose, in contrast to hemicellulose, is a semicrystalline polymer of pure glucose linked by beta-glucoside bonds. The beta linkages form linear strands that establish extensive H-bonds between them, leading to a highly stable structure that is quite resistant to degradation. Many chemical approaches have been explored to hydrolyze cellulose, though none are completely satisfactory; currently, dilute acid hydrolysis procedures are being proposed for several near-term commercialization efforts until more effective technologies are available [5].

2.1.3. Enzymatic cellulose hydrolysis. Enzymatic hydrolysis by cellulases is the ultimate goal in biomass processing for fermentation: this method has the advantages of reduced sugar loss through side reactions and it is less corrosive of process equipment [18]. In addition, the hydrolyzed product requires no neutralization prior to fermentation [19].

Cellulases consist of multicomponent enzyme complexes acting synergistically: complete cellulose hydrolysis requires the activity of an endoglucanase, which cleaves interior regions of cellulose polymers; an exoglucanase, which cleaves cellobiose units from the ends of cellulose polymers; and a beta-glucosidase, which cleaves cellobiose into its glucose subunits [20]. Because of the complexity and insolubility of the substrate, cellulase catalysis is not only relatively slow, but it is also understood much less completely than other enzymes, despite over four decades of cellulase research [8]. One of the most important organisms in the development of cellulase enzymes is Trichoderma reesei, the "ancestor of many of the most potent enzyme- producing fungi in commercial use today"[3]. By 1979, genetic enhancement had produced mutants with up to 20 times greater cellulose productivity than the original organisms found in World War II; today, surprisingly, the most lucrative cellulase market is in the manufacture of stone-washed jeans [3].

2.1.4. Fermentation. Fermentation of glucose to ethanol is performed by numerous bacteria, yeasts, and other fungi, and several yeasts have also been identified that can convert xylose to ethanol. Pentose fermentation to ethanol does not commonly co-occur with hexose fermentation to ethanol, however, spurring efforts to combine these two fermentation pathways into single organisms. Genetic engineering has since provided both bacteria and yeasts capable of fermenting both 5-carbon and 6-carbon sugars [21, 22].

Although hydrolysis of biomass cellulose by cellulases was once performed as a distinct step between pretreatment and fermentation, fermentation can begin as soon as glucose subunits are released from cellulose. The realization of this led to the development of the Simultaneous Saccharification and Fermentation (or Co-Fermentation) Process (SSF or SSCF), which now provides the great advantage of simultaneous cellulose hydrolysis and glucose fermentation [13]. This process enhances cellulase activity by relieving the product inhibition of beta-glucosidase by glucose, since the products are consumed as soon as they are produced. SSF has been patented by the Gulf Oil Company and the University of Arkansas [3]. With the availability of organisms that can ferment both pentoses and hexoses, all biomass sugars may now be simultaneously fermented in SSF/SSCF processes [23, 24]

Product inhibition of exocellulases is not eliminated, however, unless the glucose dimer cellobiose is also consumed or hydrolyzed. Since conventional S. cerevisiae strains do not metabolize cellobiose, and since cellulase preparations with sufficient beta-glucosidase activity to hydrolyze all the cellobiose are expensive to produce, much research has been directed toward the use of native cellobiose-utilizing yeast strains in SSF, either independently or in co-culture with S. cerevisiae [18].

2.1.5. Fermentation following gasification. A radically different approach to preparing biomass substrates for fermentation is found in biomass gasification. In this process, biomass is converted to synthesis gas, consisting primarily of CO, CO_2, and H_2, in addition to CH4 and N2 (25). After gasification, anaerobic bacteria such as Clostridium ljungdahlii can ferment the CO, CO_2, and H_2 into ethanol by an acetogenic process [26-29]. One advantage of the process is that, unlike acid and enzymatic hydrolysis methods, gasification can convert essentially all of the biomass, including lignin, to syngas that can be potentially fermented by bacteria [30]. Higher rates of fermentation are also achieved because the process is limited by the transfer of gas into the liquid phase instead of the rate of substrate uptake by the bacteria.

2.2. Feedstock Options

Corn and sugar cane are not long-term options for ethanol generation because of their value as foods. Exploration of various non-food forms of biomass, principally wastes, is therefore an active area of research.

Worldwide, rice straw has the greatest quantitative potential for bioethanol production, estimated at 205 gigaliters per year; this potential is concentrated in Asia, which as a region could produce up to 291 gigaliters per year of ethanol from rice straw in combination with wheat straw and corn stover. Europe has the next-largest supply of agricultural wastes, primarily in the form of wheat straw (69.2 gigaliters per year potential ethanol production); followed by North America, in which corn stover forms the majority of agricultural wastes and could supply an estimated 38.4 gigaliters per year of ethanol [11]. Bagasse, or waste derived from sugar cane, is widely available in tropical areas and is being explored by BC International Corporation (BCI), while municipal solid waste has attracted the attention of Masada Resources Group, LLC; these two companies are currently planning construction of unique biomass-to-ethanol plants [31].

Corn stover or fiber, a by-product of the corn wet-milling industry consisting of corn hulls and residual starch, is the subject of great interest as a possible substrate for ethanol production in the United States. Conversion of the starch along with the lignocellulosic components in the corn fiber could increase ethanol yields from a corn wet mill by 13 percent. In a recent study utilizing the bioethanol process development unit at the U.S. National Renewable Energy Laboratory (NREL), corn fiber was used to support continuous, integrated operation of the plant. The fiber was pretreated by high-temperature, dilute sulfuric-acid hydrolysis, and the cellulose was converted to ethanol using simultaneous saccharification and fermentation using a commercially-available cellulase and conventional Saccharomyces cerevisiae yeast that did not utilize 5-carbon sugars. Despite difficulties with bacterial contamination, which are expected to diminish with the use of recombinant, xylose- and arabinose-utilizing fermentative organisms, the attempt was successful and indicates that corn fiber could become a valuable feedstock in the United States [31]. The use of agricultural wastes is not without potential drawbacks, however. Most crop residues are currently plowed into the soil to sustain soil quality by increasing the soil organic carbon pool, enhancing activity of soil fauna, and minimizing soil erosion, and soil scientists caution that diversion of waste biomass for fuel must be undertaken cautiously [32].

Softwood forest thinnings are also being explored as potential feedstocks. Lumber manufacturing, timber harvesting, and thinning of forests to prevent wildfires generate a large quantity of softwood residues that require environmentally sound and cost-effective methods of disposal. Research in this area in the United States is currently focusing on dilute sulfuric acid hydrolysis and SO_2-steam explosion pretreatments, followed by fermentation by a Saccharomyces cerevisiae mutant yeast adapted to the inhibitory extractives and lignin degradation products present in the softwood hydrolysates [33]. Testing of recombinant xylosefermenting yeasts is also planned, and investigation is underway by Kemestrie, Inc. (Sherbrooke, QC, Canada) to identify high-value coproducts that may be derived from softwoods, focusing on antioxidants and other extractives [3].

2.3. Cellulase Engineering

The second important area in which improvement is needed for the commercialization of fuel ethanol is the conversion of lignocellulosic feedstock into the sugars to be fermented. Most current work in this area concentrates on improvement of cellulase expression, activity, and production efficiency, with the goal of reducing the cost and increasing the extent of cellulose hydrolysis.

Cellulase cost is a critical limiting factor in lignocellulose feedstock preparation. Current estimates of cellulase cost range from 3 0–50 cents per gallon of ethanol; a goal of 5 cents per gallon of ethanol is envisioned [3]. Thus, a 10-fold improvement in specific activity, production efficiency, or some combination thereof, is required.

Cellulase improvement in any of the following five critical areas could substantially improve the feasibility of bioethanol commercialization: thermostability, acid tolerance (to withstand pretreatment acidification), cellulose binding affinity, specific activity, and reduced nonspecific binding to lignin [14]. While these features are theoretically approachable by genetic engineering techniques, use of these techniques is presently limited by the incomplete understanding of cellulase catalysis. A primary reason for this is that cellulose-cellulase systems involve soluble enzymes working on insoluble substrates, which represents a substantial increase in complexity from homogeneous enzyme-substrate systems. In addition, the catalytic system involves the synergistic activities of three different enzymes [3]. Still, a number of promising avenues are currently being explored.

2.3.1. Cellulase component engineering. The most fundamental improvements that are needed are within the cellulase components themselves, these are the endoglucanases, exoglucanases, and cellobiohydrolases (or beta-glucosidases). Using Trichoderma, Clostridium, Cellulomonas, and Thermobifida, among others, efforts are underway to improve activity, expression, and specificity of these components through site-directed mutagenesis, use of heterologous promoters to direct transcription, and modeling to reveal structure-function relationships [34–39].

One of the most active cellulase components known is the endoglucanase E1 from Acidothermus cellulolyticus. Two leading industrial enzyme producers, Novozymes (www.novozymes.com) and Genencor International (www.genencor.com), are currently contributing to the cellulase improvement effort with support from DOE. In 1998, J. Sakon and colleagues at the University of Arkansas showed that performance of a ternary system was improved 13 percent by site-directed modification of one active site amino acid in Acidothermus E1; currently they are pursuing E1 mutations that modify the biomass interactive surface.

2.3.2. Chimeric cellulase systems. Cellulase components from diverse organisms, primarily bacteria and fungi, are being combined in ways that yield overall improved activity. Baker and colleagues have successfully combined bacterial and fungal cellulases in vitro [40], showing that these mixtures can be competitive with a native ternary system from T. reesei [41]. Work with expansins, proteins that enable extension of plant cell walls during plant cell growth, has also shown enhancement of hydrolysis of microcrystalline cellulose in a mixed Trichoderma cellulase preparation [42]. The initial approaches to developing artificial cellulase systems, still instructive after nine years, are reviewed in [43].

2.3.3. Heterologous expression. The next logical step in chimeric cellulase systems is the cloning of cellulases from one organism into another; this avenue is being explored as well, as shown by the expression of the T. reesei cellobiohydrolase I in Pichia pastoris [44]. Especially important for commercial production, cellulases are being expressed in plants such as tobacco and potatoes, potentially providing more abundant sources of the enzymes [45].

Another innovative approach to the heterologous expression of cellulases is the expression of heat-activated cellulases within biomass crops themselves, with the idea that the plants grow normally until harvested and exposed to elevated temperatures, at which point heat- activated cellulases hydrolyze the cellulose without need for externally added enzymes [46].

2.3.4. Cellulase performance assays. Convenient, accurate, efficient assays are central to the development of any new technology. The diafiltration saccharification assay (DSA) developed at the NREL produces precise, detailed progress curves for enzymatic saccharification of cellulosic materials under conditions that mimic those of SSF. From this method, it is possible to describe the performance of a given cellulase preparation over a wide range of loading and reaction times with comparatively little data [47, 48, 3].

2.3.5. Proteomic analysis, microarray analysis, and modeling. Proteomics is an emerging set of techniques that has proven extremely useful in understanding the interactions of multienzyme systems. Hydrolysis of complex organic substrates is an ideal candidate for proteomic analysis, as it involves a number of enzymes: $\beta3$-1,4-endoglucanases, $\beta3$-1,4-cellobiohydrolases, xylanases, $\beta3$-glucosidases, α -L-arabinofuranosidase, acetyl xylan esterase, $\beta3$- mannanase, and α-glucuronidase in T. reesei, for example. At the NREL, the expression of these enzymes is being investigated under various conditions by proteomic methods and compared to corresponding enzyme activities using the DSA assay [3].

To reveal gene expression responses to environmental conditions in both wild-type and genetically engineered microbes, microarray analysis is underway and could become a valuable industrial tool for evaluation of new recombinant organisms [49]. Mathematical molecular analysis is also being employed to gain greater understanding of structure-function relationships to complement the physiological understanding provided by proteomic and microarray analysis. Current work includes molecular mechanics efforts by Brady and colleagues at Cornell University as well as Palma and colleagues; cellulase crystallization work is also in progress by a number of groups [50-53].

2.4. Fermentation Technologies

While fermentation of glucose into ethanol is a well-understood process that occurs widely among microorganisms, the fermentation of pentoses such as xylose, which are abundant in biomass, has posed significant challenges. Recently, however, this challenge has been addressed by creating recombinant yeasts and bacteria [21, 22], although the solution may not yet have been fully optimized.

A second important aspect limiting commercialization is the fermentation efficiency of the microorganisms: in typical fermentation pathways for glucose and xylose to ethanol, one contributor to the high cost of ethanol production is the loss of half of the fixed carbon to products other than ethanol [5].

The ideal bioethanol-fermenting microorganism would therefore readily ferment all biomass sugars, resist toxic effects of aromatic lignin subunits and other inhibitory byproducts such as acetate, be thermostable and acid-tolerant, and produce a highly active cellulase multienzyme complex [54].

2.4.1. Pentose fermentation. As mentioned above, Escherichia, Klebsiella, and Zymomonas have now been engineered to ferment not only glucose but also xylose and arabinose sugars [5, 54, 55]. Some of these are already experiencing commercial use as well: BC International Corporation (www.bcintl-corp.com) is using genetically engineered Escherichia coli to produce ethanol from biomass sugars, and Arkenol Inc. (www.arkenol.com) is using Zymomonas in its concentrated-acid process.

In another example, Zymomonas mobilis has been transformed with Escherichia coli xylose isomerase, xylulokinase, transaldolase, and transketolase genes. Expression of the added genes are under the control of Zymomonas mobilis promoters. This genetically modified microorganism, patented by the Midwest Research Institute, is now able to ferment mixtures of xylose, arabinose, and glucose to produce ethanol [56, 57].

2.4.2. Combined cellulolysis and fermentation. Consolidated bioprocessing (CBP), in which the production of cellulolytic enzymes, hydrolysis of biomass, and fermentation of resulting sugars to desired products occur in one step, is currently envisioned as the most promising and eminently achievable path toward optimally efficient bioethanol production [54]. Efforts to develop such a culture through engineering fermentative capacity into cellulolytic organisms, as well as the alternative, engineering cellulolytic capacity into fermentative organisms, are both underway and have been reviewed extensively [54].

In one example, Ingram and colleagues cloned two Erwinia endoglucanase genes into an ethanol-producing Klebsiella species, producing a new microbe that produced up to 22 percent more ethanol when fermenting crystalline cellulose synergistically with added fungal cellulases [58]. Cellulase genes have also been introduced into Lactobacillus, although not necessarily for biomass utilization [59], and cellobiose utilization capability has been engineered into Saccharomyces cerevisiae [60]. In an alternative example, with the additional goal of offering improved relief from product inhibition in SSF, in which cellobiose inhibition of exoglucanase is problematic, ethanol-producing genes have been successfully introduced into native cellobioseutilizing bacteria [61, 62].

2.4.3. Synergistic co-cultures. In experiments involving the cellobio se-fermenting recombinant, Klebsiella oxytoca P2, in co-cultures with ethanol-tolerant strains of Saccharomyces pastorianus, Kluyveromyces marxianus, and Zymomonas mobilis, the combinations produced more ethanol, more rapidly, than any of the constituent strains. This was accomplished by early ethanol production by K. oxytoca, while ethanol produced in the later stages was primarily by the more ethanol-tolerant strain [18].

2.4.4. Improved thermotolerance. Ethanol fermentation at elevated temperatures (>55°C) would facilitate product recovery, but thermophilic bacteria are poor ethanol producers. In addition, thermophilic Clostridium and Thermoanaerobium species have been investigated for potential as ethanol producers, but were consistently limited by end-product inhibition and solvent-induced membrane damage [63].

In addition, efforts are underway to eliminate acid production during fermentation through genetic engineering, enabling use of salt-intolerant thermophilic strains like Thermoanaerobacterium thermosaccharolyticum, a microbe tolerant to high levels of ethanol but intolerant of salt accumulation during pH-controlled fermentations. If such a thermotolerant organism could be improved further to produce high-activity cellulases, a highly productive, anaerobic, ethanol-producing strain could result. Cellulase production could, however, pose an insurmountable energy burden to a fermentative organism; the energetic considerations of this combination are being evaluated [5].

2.4.5. Fermentation of synthesis gas. Rajagopalan and coworkers report the discovery of a clostridial bacterium, P7, that converts mixtures of CO, CO_2, and N2 into ethanol, butanol and acetic acid, with high ethanol production and selectivity compared to previous isolates. The authors report process parameters and consider options for improving ethanol yield [64].

2.5. Coproduct Development

Finally, the investigation of potential ethanol coproducts is underway. Biomass sugars can support the production of many other products along with ethanol, including organic acids and other organic alcohols, 1 ,2-propanediol, and aromatic chemical intermediates. If these coproducts were sufficiently valuable, they could help greatly offset costs of ethanol production. However, such coproducts must be chosen carefully to ensure that sufficient markets are available [65].

Additional coproducts may be available from lignin: this material is present at 15–30 percent by weight in all lignocellulosic biomass, and any bioethanol production process will have lignin as a residue. A team of researchers from the NREL, the University of Utah, and Sandia National Laboratories is working to develop a process for making oxygenate fuel additives from lignin; these processes are chemical in nature and are detailed in other materials referenced [66- 68].

3. Research Priorities

The consensus among researchers and supporters of bioethanol research, in addition to those engaged in commercial projects, is that the improvement of cellulase enzyme activity and cellulase production, both to increase the efficiency of release of fermentable sugars from biomass and to reduce cellulase cost, are two of the greatest advances needed in the effort to commercialize fuel ethanol production [19, 5]. In addition is the development of enzymatic pretreatment processes to release lignin from carbohydrate components [42, 9] and further improvement of fermentative organisms [69, 64], with the particular goal of designing microbes capable of consolidated bioprocessing [54].

4. Commercialization

4.1. Cellulases

Although many commercial preparations of cellulase exist, costs have remained high because present applications are in higher-value markets (food and clothing) than fuels. In addition, these applications typically require much less than 100 percent cellulose hydrolysis, in contrast to ethanol production; much improvement is therefore needed to advance the current cellulase enzyme industry to the point at which it can support the fuel ethanol industry. As a result, many fuel ethanol commercialization efforts are choosing to use acid hydrolysis techniques for cellulose hydrolysis until cellulase preparations become less expensive or until recombinant microbes capable of combined cellulolysis and fermentation are perfected [3]. Toward this goal, the DOE biofuels program is working with the two largest global enzyme producers, Genencor International and Novozymes Biotech Incorporated, to achieve a 10-fold reduction in cost of cellulases [3]

4.2. Ethanol Plants

The first dedicated large-scale plants for the conversion of waste biomass to ethanol are now in planning and/or construction phases by BCI and the Masada Resource Group (www.masada.com), while Iogen Corporation (www.iogen.com) is currently operating a 50 ton per week pilot plant. BCI and the DOE Office of Fuels Development have formed a cost-shared partnership to develop a biomass-to-ethanol plant intended to produce 20 million liters of ethanol per year initially from sugar cane waste (bagasse) and other biomass, utilizing an existing ethanol plant in Jennings, LA. Dilute acid hydrolysis will be used to recover sugar from bagasse initially, allowing for addition of enzyme hydrolysis when cellulases become less expensive. A proprietary genetically-engineered microbe will ferment the sugars to ethanol. BCI is also planning to operate a plant in Gridley, CA, in which cellulases will be used in conversion of commercial rice straw to ethanol, again with partial DOE support.

The Masada plant is expected to produce 9.5 million gallons from municipal solid waste using Masada's patented CES OxyNolTM concentrated acid hydrolysis technology in New York [3].

Petro-Canada, the second largest petroleum refining company in Canada, began to co-fund research and development on biomass-to-ethanol technology with Iogen in 1997. Petro-Canada, Iogen, and the Canadian government then began plans to fund construction of a demonstration plant based on Iogen's cellulase enzyme technology in an SSF process [3]. The plant of Iogen, a leading producer of cellulases, has completed a 40 ton per day biomass-to-ethanol demonstration facility that is now in its start-up phase [5].

In the pulp and paper industry, Tembec and Georgia Pacific are using dilute acid hydrolysis to dissolve hemicellulose and lignin from wood, producing a cellulose pulp that can be fermented to ethanol. The lignin is then used to generate energy, through combustion, or converted to other products such as concrete additives and soil stabilizers [3].

Pursuing the gasification and syngas-to-ethanol fermentation, BioEngineering Resources, Inc. (BRI) has developed syngas technology to the extent that plans are underway to pilot the technology as a first step toward commercialization. BRI has developed bioreactor systems for fermentation that result in retention times of minutes or less, yielding low equipment costs [3].

 John G. Taylor

REFERENCES

[1] Turner, J. (1999). *A realizable renewable energy future*, Science 285:687-689.

[2] United States Department of Energy (2003). *Biofuels for Sustainable Transportation.* Office of Energy Efficiency and Renewable Energy, http://www.ott.doe.gov/biofuels/advanced_bioethanol.html.

[3] United States Department of Energy (2003). *Advanced Bioethanol Technology*, Office of Energy Efficiency and Renewable Energy, http://www.ott.doe.gov/biofuels/advanced bioethanol.html.

[4] Kaltschmitt, M., G. Reinhardt, and T. Stelzer (1997). *Life cycle analysis of biofuels under different environmental aspects*, Biomass Bioenergy 12:121-134

[5] Mielenz (2001). *Ethanol production from biomass: technology and commercialization status*, Current Opinion in Microbiology 4:324-329.

[6] United States Department of Energy (2004). *Energy Security and the U.S. Transportation Sector*, Office of Energy Efficiency and Renewable Energy, Biomass Program, http://www.eere.energy.gov/biomass/national_energy_security.html#security.

[7] United States Department of Energy (2004). *Alternative Fuels Data Center*, Office of Energy Efficiency and Renewable Energy, Clean Cities Program, http://www.eere.energy.gov/cleancities/afdc/infrastructure/station_counts.html.

[8] Sheehan, J. (2001). *The road to bioethanol: a strategic perspective of the US Department of Energy's national ethanol program*, in Glycosyl Hydrolases for Biomass Conversion (M. Himmel, J. O. Baker, and J. Saddler, Eds.). American Chemical Society, Washington, D. C., pp 2-25.

[9] Sun, Y., and J. Cheng (2002). *Hydrolysis of lignocellulosic materials for ethanol production: a review*, Bioresource Technol 83:1-11.

[10] Goldemburg, J., S. Coelho, P. Nastari, and O. Lucon (2004). *Ethanol learning curve -- the Brazilian experience*, Biomass Bioenergy 26:301-304.

[11] Kim, S., and B. Dale (2004). *Global potential bioethanol production from wasted crops and crop residues*, Biomass Bioenergy 26:361-375.

[12] Himmel, M. E., S. A. Adney, J. O. Baker, R. Elander, J. D. McMillan, R. A. Nieves, J. J. Sheehan, S. R. Thomas, T. B. Vinzant, and M. Zhang (1997). *Advanced Bioethanol Production Technologies: A Perspective*, *in* Fuels and Chemicals from Biomass. American Chemical Society, Washington, DC, pp 1-45.

[13] Wyman, C. E. (1999). *Biomass ethanol: Technical progress, opportunities, and commercial challenges*, Annu Rev Energy Environ 24:189-226.

[14] Wooley, R., M. Ruth, D. Glassner, and J. Sheehan (1999). *Process design and costing of bioethanol technology: A tool for determining the status and direction of research and development*, Biotechnol Prog 15:794-803.

[15] Mosier, N., C. E. Wyman, B. Dale, R. Elander, Y. Y. Lee, M. Holzapple, and M. Ladisch (2005). *Features of promising technologies for pretreatment of lignocellulosic biomass*, Bioresource Technol 96:673-686.

[16] Amen-Chen, C., H. Pakdel, and C. Roy (2001). *Production of monomeric phenols by thermochemical conversion of biomass: A review*, Bioresource Technol 79:277-299.

[17] United States Department of Energy (2004). Pretreatment Technology Evaluation, Office of Energy Efficiency and Renewable Energy, Biomass Program, http://www.eere.energy.gov/biomass/technology_evaluation.html.

[18] Golias, H., G. J. Dumsday, G. A. Stanley, and N. B. Pamment (2002). *Evaluation of a recombinant Klebsiella oxytoca strain for ethanol production from cellulose by simultaneous saccharification and fermentation: comparison with native cellobiose-utilising yeast strains and performance in co-culture with therm otolerant yeast and Zymomonas mobilis,* J Biotechnol 96:155-168.

[19] Himmel, M. E., J. O. Baker, and J. N. Saddler (2001). *Glycosyl hydrolases for biomass conversion.* American Chemical Society : Distributed by Oxford University Press, Washington, DC.

[20] Himmel, M. E., W. S. Adney, J. O. Baker, R. A. Nieves, and S. R. Thomas (1996). *Cellulases: Structure, Function, and Application, in* Handbook on Bioethanol (C. E. Wyman, Ed.). Taylor & Francis, Washington, DC, pp 143-161.

[21] Dien, B., R. Hespell, H. Wyckoff, and R. Bothast (1998). *Fermentation of hexose and pentose sugars using a novel ethanologenic* Escherichia coli strain, Enzyme Microb Technol 23:366-371.

[22] Toivari, M., A. Aristidou, L. Ruohonen, and M. Penttila (2001). *Conversion of xylose to ethanol by recombinant* Saccharomyces cerevisiae: *Importance of xylulokinase (XKS1) and oxygen availability,* Metabolic Eng 3:236-249.

[23] McMillan, J. D., M. M. Newman, D. W. Templeton, and A. Mohagheghi (1999). *Simultaneous saccharification and cofermentation of dilute-acid pretreated yellow poplar hardwood to ethanol using xylose-fermenting* Zymomonas mobilis, Appl Biochem Biotechnol 77/79:649-665.

[24] Teixeira, L. C., J. C. Linden, and H. A. Schroeder (2000). *Simultaneous saccharification and cofermentation of peracetic acid-pretreated biomass,* Appl Biochem Biotechnol 84/86:111-127.

[25] McKendry, P. (2002). *Energy production from biomass. Part 3: Gasification technologies,* Bioresource Technology 83:55-63.

[26] Ljungdahl, L. G. (1986). *The autotrophic pathway of acetate synthesis in acetogenic bacteria,* Annu Rev Microbiol 40:415-450.

[27] Wood, H. G., S. W. Ragsdale, and E. Pezacka (1986). *The acetyl-CoA pathway of autotrophic growth,* FEMS Microbiol Rev 39:345-362.

[28] Abrini, A., H. Naveau, and E. J. Nyns (1994). Clostridium autoethanogenum, *sp. nov., an anaerobic bacterium that produces ethanol from carbon monoxide,* Arch Microbiol 161 :345-351.

[29] Phillips, J. R., E. C. Clausen, and J. L. Gaddy (1994). *Synthesis gas as substrate for the biological production of fuels and chemicals,* Appl Biochem Biotechnol 45/46:145-154.

[30] Reed, T. B., and D. Jantzen (1979). *Biomass gasification: Principles and technology,* Energy Technology Review 67:27-90.

[31] Schell, D. J., C. J. Riley, N. Dowe, J. Farmer, K. N. Ibsen, M. F. Ruth, S. T. Toon, and R. E. Lumpkin (2004). *A bioethanol process development unit: initial operating experiences and results with a corn fiber feedstock,* Bioresource Technol 91:179-188.

[32] Lal, R. (2005). *World crop residues production and implications of its use as a biofuel,* Environ Int 3 1:575-584.

[33] Keller, F. A., D. Bates, R. Ruiz, and Q. Nguyen (1998). *Yeast adaptation on softwood prehydrolysate*, Appl Biochem Biotechnol 70:137-148.

[34] Boer, H., T.T. Teeri, and A. Koivula (2000). *Characterization of* Trichoderma reesei *cellobiohydrolase Cel7A secreted from* Pichia pastoris *using two different promoters*, Biotechnol Bioeng 69:486-494.

[35] Boisset, C., C. Fraschini, M. Schulein, B. Henrissat, and H. Chanzy (2000). *Imaging the enzymatic digestion of bacterial cellulose ribbons reveals the endo character of the cellobiohydrolase Cel6A from* Humicola insolens *and its mode of synergy with cellobiohydrolase Cel7A*, Appl Environ Microbiol 66:1444-1452.

[36] Brun, E., P.E. Johnson, A.L. Creagh, P. Tomme, P. Webster, C.A. Haynes, and L.P. McIntosh (2000). *Structure and binding specificity of the second N-terminal cellulose-binding domain from* Cellulomonas fimi *endoglucanase C*, Biochem 39:2445-2458.

[37] Irwin, D. C., S. Zhang, and D. B. Wilson (2000). *Cloning, expression and characterization of a family 48 exocellulase, Cel48A, from* Thermobifida fusca, Eur J Biochem 267:4988-4997.

[38] Quixley, K. W., Reid S.J. (2000). *Construction of a reporter gene vector for* Clostridium beijerinckii *using a Clostridium endoglucanase gene*, J Mol Microbiol Biotechnol 2:53- 57.

[39] Zhang, S., D. C. Irwin, and D. B. Wilson (2000). *Site-directed mutation of noncatalytic residues of* Thermobifida fusca *exocellulase Cel6B*, Eur J Biochem 267:3101-3115.

[40] Baker, J. O., W. S. Adney, R. A. Nieves, S. R. Thomas, and M. E. Himmel (1995). *Synergism in Binary Mixtures of Bacterial and Fungal Cellulases: Endo/Exo, Exo/Exo, and Endo/Endo Interactions, in* Enzymatic Degradation of Insoluble Polysaccharides (J.

[41] N. Saddler and M. H. Penner, Eds.). American Chemical Society, Washington, DC, pp 113- 141.

[42] Baker, J. O., C. I. Ehrman, W. S. Adney, S. R. Thomas, and M. E. Himmel (1998). *Hydrolysis of cellulose using ternary mixtures of purified cellulases*, Appl Biochem Biotechnol 70:395-403.

[43] Baker, J. O., M. R. King, W. S. Adney, S. R. Decker, T. B. Vinzant, S. L. Lantz, R. E. Nieves, S. R. Thomas, L. -C. Li, D. J. Cosgrove, and M. E. Himmel (2000). *Investigation of the cell wall loosening protein expansion as a possible additive in the enzymatic saccharification of lignocellulosic biomass*, Appl Biochem Biotechnol. 84-86:217-223.

[44] Thomas, S. R., R. A. Laymon, Y. C. Chou, M. P. Tucker, T. B. Vinzant, W. S. Adney, J. O. Baker, R. A. Nieves, J. R. Mielenz, and M. E. Himmel (1995). *Initial Approaches to Artificial Cellulase Systems for Conversion of Biomass to Ethanol, in* Enzymatic Degradation of Insoluble Polysaccharides (J. N. Saddler and M. H. Pennery, Eds.). American Chemical Society, Washington, DC, pp 208-23 6.

[45] Godbole, S., S. R. Decker, R. A. Nieves, W. S. Adney, T. B. Vinzant, J. O. Baker, S. R. Thomas, and M. E. Himmel (1999). *Cloning and expression of* Trichoderma reesei *CBH I in Pichia pastoris*, Biotechnology Progress 15:828-833.

[46] Dai, Z., B. S. Hooker, D. B. Anderson, and S. R. Thomas (2000). *Expression of* Acidothermus cellulolyticus *endoglucanase E1 in transgenic tobacco: Biochemical characteristics and physiological effects*, Transgenic Research 9:43-54.

[47] Office of News Services (2001). New CU-Boulder Research May Reduce Renewable Fuels Costs, B. University of Colorado, http://www.colorado.edu/PublicRelations/ NewsReleases/2001/1 244.html.

[48] Baker, J. O., T. B. Vinzant, C. I. Ehrman, W. S. Adney, and M. E. Himmel (1997) *A membrane-reactor saccharification assay to evaluate the performance of cellulases under simulated SSF conditions*, Appl Biochem Biotechnol 63:585-595.

[49] Vinzant, T. B., C. I. Ehrman, and M. E. Himmel (1997). *SSF of pretreated hardwoods. Effect of native lignin content*, Appl Biochem Biotechnol 62:97-102.

[50] Tao, H., R. Gonzalez, A. Martinez, M. Rodriguez, L. Ingram, J. Preston, and K. Shanmugam (2001). *Engineering a homoethanol pathway in* Escherichia coli. *Increased glycolytic flux and levels of expression of glycolytic genes during xylose fermentation*, J Bacteriol 185:2979-2988.

[51] Sakon, J., W. Adney, M. Himmel, S. Thomas, and P. Karplus (1996). *Crystal structure of thermostable Family 5 endocellulase EI from* Acidothermus cellulolyticus *in complex with cellotetraose*, Biochemistry 35:10648-10660.

[52] Shirai, T., H. Ishida, J. Noda, T. Yamane, K. Ozaki, Y. Hakamada, and S. Ito (2001). *Crystal structure of alkaline cellulase K. Insight into the alkaline adaptation of an industrial enzyme*, J Mol Biol 3 10:1079-1087.

[53] Hirvonen, M., and A. C. Papageorgiou (2003). *Crystal structure of a family 45 endoglucanase from* Melanocarpus albomyces. *Mechanistic implications based on the free and cellobiose-bound forms*, J Mol Biol 329:403 -410.

[54] Mandelman, D., A. Belaich, J. P. Belaich, N. Aghajari, H. Driguez, and R. Haser (2003). *X-Ray crystal structure of the multidomain endoglucanase Cel9G from* Clostridium cellulolyticum *complexed with natural and synthetic cello-oligosaccharides*, J Bacteriol 185:4127–4135.

[55] Lynd, L. R., P. J. Weimer, W. H. van Zyl, and I. S. Pretorius (2002). *Microbial cellulose utilization. Fundamentals and biotechnology*, Microbiol Molec Biol Rev 66:506-577.

[56] Dien, B. S., M. A. Cotta, and T. W. Jeffries (2003). *Bacteria engineered for fuel ethanol production. Current status*, Appl Microbiol Biotechnol 63:258-266.

[57] Picataggio, S. K., M. Zhang, C. K. Eddy, K. A. Deanda, and M. Finkelstein (1996). *Recombinant Zymomonas for pentose fermentation*, U.S. Patent 5514583.

[58] Zhang, M., Y.-C. Chou, S. K. Picataggio, and M. Finkelstein (1998). *Single Zymomonas* mobilis *strain for xylose and arabinose fermentation*, U.S. Patent 5843760.

[59] Zhou, S., F. C. Davis, and L. O. Ingram (2001). *Gene integration and expression and extracellular secretion of* Erwinia chrysanthemi *endoglucanase CelY (celY) and CelZ (celZ) in ethanologenic* Klebsiella oxytoca *P2*, Appl Environ Microbiol 67:6-14.

[60] Cho, J., Y. Choi, and D. Chung (2000). *Expression of* Clostridium thermocellum *endoglucanase gene in* Lactobacillus gasseri *and* Lactobacillus johnsonii *and characterization of the genetically modified probiotic lactobacilli*, Curr Microbiol 40:257-263.

[61] McBride, J. E., J. J. Zietsman, W. H. van Zyl, and L. R. Lynd (2005). *Utilization of cellobiose by recombinant ß-glucosidase-expressing strains of Saccharomyces cerevisiae: Characterization and evaluation of the sufficiency of expression*, Enzyme Microb Technol 37:93-101.

[62] Wood, B. E., and L. O. Ingram (1992). *Ethanol production from cellobiose, amorphous cellulose, and crystalline cellulose by recombinant* Klebsiella oxytoca *containing*

chromosomally integrated Zymomonas mobilis *genes for ethanol production and plasmids expressing thermostable cellulase genes from* Clostridium thermocellum, Appl Environ Microbiol 58:2103-2110.

[63] Moniruzzaman, M., X. Lai, S. W. York, and L. O. Ingram (1997). *Extracellular melibiose and fructose are intermediates in raffinose catabolism during fermentation to ethanol by engineered enteric bacteria,* J Bacteriol 179:1880-1886.

[64] Ingram, L. (1990). *Ethanol tolerance in bacteria,* CRC Crit Rev Biotechnol 9:305-319.

[65] Rajagopalan, S., R. P. Datar, and R. S. Lewis (2002). *Formation of ethanol from carbon monoxide via a new microbial catalyst,* Biomass and Bioenergy 23 :487-493.

[66] Lynd, L., C. Wyman, and T. Gerngross (1999). *Biocommodity engineering,* Biotechnol Prog 15:777-793.

[67] Miller, J. E., L. Evans, A. Littlewolf, and D. E. Trudell (1999). *Batch microreactor studies of lignin and lignin model compound depolymerization by bases in alcohol solvents,* Fuel 78:1363-1366.

[68] Shabtai, J., and et al. (1999). *Process for conversion of lignin to reformulated hydrocarbon gasoline,* U.S. Patent 5,959,167.

[69] Shabtai, J. S., W. Zmierczak, E. Chornet, and D. K. Johnson (1999). *Lignin conversion to high-octane fuel additives,* Fourth Biomass Conference of the Americas, Oakland, CA.

[70] Wyman, C. E. (2001). *Twenty years of trials, tribulations, and research progress in bioethanol technology: Selected key events along the way,* Appl Biochem Biotechnol 91-93 :5-21.

B. Biodiesel

1. Introduction

Among biofuels (biomass, biohydrogen, bioethanol, and biodiesel), biodiesel is currently the most fully developed and widely used, with many countries producing in excess of 100,000 tons per year, including Belgium, France, Germany, Italy, and the United States. Japan is also a leading entity in biodiesel research and in production of biodiesel from waste oils [1, 2]. The current status of biodiesel is understandable given its many convenient features, such as the following: it can be used in common compression-injection engines (CIE), it poses no particular difficulties compared to fossil fuels in handling, transport, or storage, and it can be mixed in any proportion with conventional diesel fuel. Because it is oxygenated and typically lacks sulfur contamination, it also burns quite cleanly, greatly reducing the output of ash, particulate matter, CO, and sulfur oxides in comparison to conventional diesel fuels [1]

The primary obstacle to its more widespread adoption is simply its cost—to date, the raw material plant oils (typically rapeseed, soybean, canola, etc.) have been much more expensive than petroleum [3]. Diminishing accessibility of petroleum, however, as well as new technology allowing recycling of waste vegetable oils into biodiesel, are eroding this obstacle, with the result that biodiesel is emerging as one of the most promising examples of the use of biotechnology for economic sustainability and pollution prevention [4].

1.1. History and Development

Conventional diesel fuel is the portion of crude oil that is distilled between approximately 200°C (392°F) and 370°C (698°F), higher than the boiling range of gasoline and consistent with its heavier, oilier composition. Diesel fuel is ignited in a CIE cylinder by the heat of air under high compression, in contrast to motor gasoline, which is ignited by an electrical spark. Because of the mode of ignition, a high cetane number (cetane is the hydrocarbon C16H34, or 1 -hexadecane, that ignites very easily under compression and is therefore used as a standard in determining diesel fuel ignition performance) is required in a good diesel fuel. Two grades of diesel fuel have been established by the ASTM: Diesel 1 and Diesel 2. Diesel 1 is a kerosene- type fuel, which is lighter, more volatile, and cleaner than Diesel 2, and is used in engine applications with more frequent changes in speed and load. Diesel 2 is used in industrial and heavy mobile service [5, 6]. The term Biodiesel, in turn, is collective describing fuels comprised of esterified plant oils or animal fats. These biological lipids originate as mixtures of triglycerides and free fatty acids that are derivatized through transesterification (also known as alcoholysis) with acid, base, or enzymatic catalysis to form, most commonly, methyl or ethyl esters (Figure 14) [7, 1, 3].

Before the advent of biodiesel, biological oils were investigated extensively in their native forms for use in diesel engines. Indeed, Rudolf Diesel himself experimented with vegetable oils in his engine (Figure 15) over 100 years ago [7]. Unfortunately, however, plant oils and especially animal fats typically have sufficiently high viscosities, and sufficiently low cetane numbers, flash points, and combustibility, that they show numerous undesirable properties in CIEs. Principal among these are carbon deposition on engine parts, gelling and polymerization during storage and in cool temperatures, and contamination of engine lubricating oils leading to deterioration of lubricant and ultimately engine performance [8, 7].

As a result, the availability of inexpensive petroleum led quickly to the nearly exclusive use of diesel fuel, which lacked the undesirable performance properties but brought with it its own undesirable, if not initially appreciated, environmental consequences. Nevertheless, experimentation with soybean, canola, and other vegetable oils, as well as beef tallow and other animal fats, continued for many decades as researchers attempted to discover mixtures with petroleum fuels that minimized the problems of incomplete combustion posed by the biolipids. While these measures successfully diminished the rate of engine and lubricating oil deterioration, most problems persisted to some extent, and at present the use of and experimentation with unmodified biolipids has effectively ceased [7].

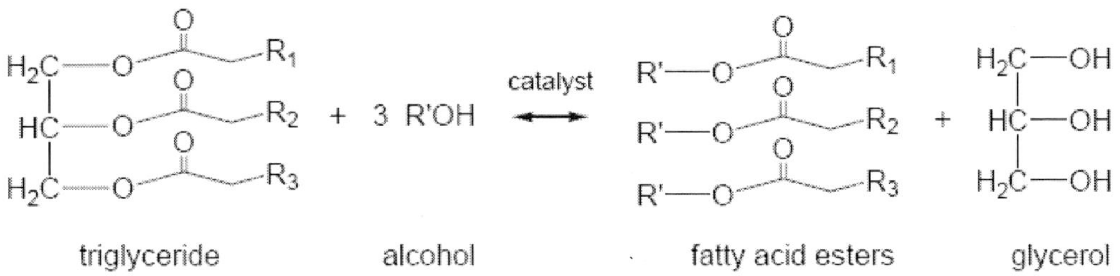

Figure 14. Transesterification of a triglyceride with an alcohol, showing the production of fatty acid esters and glycerol. Adapted from [1].

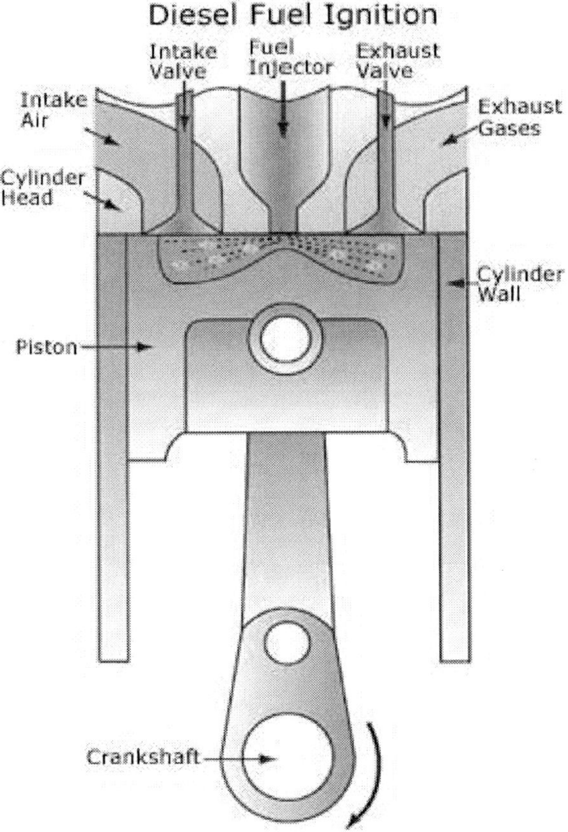

Figure 15. The diesel engine, invented by Rudolf Diesel in 1894 and characterized by its ignition of hydrocarbon fuels by compression with heated air, rather than by ignition with an electrical spark [41].

2. State of the Science

2.1. Feedstocks and Coproducts

An important challenge faced by biodiesel in competing with petroleum diesel is the comparatively high price of its plant oil feedstocks. In June, 2004, diesel prices ranged from $1.50 per gallon in Oklahoma to $2.20 per gallon in Seattle, Washington [9], while the cost of soybean oil, the primary biodiesel feedstock in the United States, was reported at $2.19 per gallon (27.34 cents per lpound) before conversion to biodiesel even occurred [10].

The margin is narrowing, however (in May 2005, diesel prices ranged from $1.98 per gallon in Knoxville, Tennessee to $2.53 per gallon in Seattle, Washington [11], while the cost of soybean oil dropped to 22.31 cents per pound [12]), and on April 25, 2005, Blue Sun 100 percent pure biodiesel (B100) reached a low of $2.39 per gallon in Denver, Colorado, compared to local petroleum diesel prices of $2.29 per gallon (www.boulderbiodiesel.org).

In addition, tax incentives are becoming popular internationally as governments attempt to reduce their dependence on foreign oil. Even in the United States, which has traditionally subsidized petroleum use substantially in comparison to Europe (diesel fuel cost is greater by a factor of 1.5–2.5 across Europe [13]), bills have been recently introduced to extend the federal biodiesel excise tax credit that provides one cent credit per percent of biodiesel per gallon. Under this incentive, taken by the producer and passed on to the consumer, B100 therefore realizes a $1.00 per gallon-cost reduction, and the price of 20 percent biodiesel combined with 80 percent petroleum diesel (B20) becomes comparable to petroleum diesel [14]. In other countries, biodiesel production and use is facilitated by higher petroleum prices (diesel fuel cost is greater by a factor of 1.5-2.5 across Europe [13]), by use of less-expensive feedstocks, and by a variety of creative pro-biodiesel incentives. In Europe, biodiesel production has been encouraged by EU farm production programs: much of the biodiesel expansion in the 1990s occurred as a result of EU policies that allowed farmers to grow crops for industrial uses, including oilseeds, on set-aside land. Tax benefits from Germany, Austria, and France have also encouraged biodiesel production and use, allowing biodiesel to find far greater success in these countries than is possible in the United States with its lack of government incentives [15]. Rapeseed oil, the predominant feedstock in Europe, is also priced lower than soybean oil [16].

Nevertheless, the cost of feedstock oils is expected to continue to be a concern in biodiesel success, promoting investigation of alternative oils. Waste cooking oil, tallow, and lard are quite inexpensive feedstocks, for example, that are currently used for biodiesel production in Japan and are promising as well for other areas of Asia that have limited agricultural land and where vegetable oils are fairly expensive [7, 17].

Another approach is the development of processes that yield valuable coproducts: glycerol is a clear candidate coproduct in biodiesel production, especially if the transesterification of plant oils is accomplished enzymatically. Enzyme use eliminates the need for an alcohol evaporation process, necessary for glycerol recovery from alkali transesterifications, and also minimizes saponification (soap formation) of glycerol and it attendant purification difficulties. Because of the favorable commodity market for glycerol, the cost of biodiesel could be lowered significantly if biodiesel plants incorporated glycerol recovery facilities [7] or provisions for production of other high-value coproducts, such as caproic or propionic acids [15].

2.2. Abiotic Processing

2.2.1. Pyrolysis. Pyrolysis, the use of heat or heat plus a catalyst in the absence of oxygen to convert one substance into another, has been investigated worldwide for much of the last century for biofuel production. In this process, triglyceride fatty acids are "cracked" from the glycerol backbone and further decomposed, yielding a mixture comprised primarily of alkanes and alkenes with smaller proportions of carboxylic acids and aromatics. While this approach effectively diminishes the viscosity of the oils and fats, a number of new problems arise: the process tends to generate a greater proportion of lower molecular-weight products (gasoline), the equipment required is expensive for modest throughputs, and the removal of oxygen during the thermal processing also removes the environmental benefits of using an oxygenated fuel [1, 7].

2.2.2. Microemulsions. Another approach to minimize vegetable oil viscosity is the formation of water-oil microemulsions, in which an oil is stably dispersed in a solvent such as methanol, ethanol, or butanol in 1-150 nanometer micelles by association with ionic or nonionic amphiphiles. While viscosity has been successfully diminished, and spray performance has been successfully enhanced by this technique, the problems of incomplete combustion leading to carbon deposition, clogging, and sticking of moving parts inherent to use of unmodified vegetable oils persisted [1, 3, 7].

2.2.3. Transesterification. The transesterification of triglyceride fatty acids with alcohols, yielding esters (Figure 14), has emerged as the technology that yields products most similar to conventional diesel fuel in chemical and combustion characteristics (Table 3). This is a stepwise process, esterifying one fatty acid at a time, that ultimately yields fatty acid esters as well as glycerol. Because each step is reversible, the product yield is enhanced greatly by providing an excess of alcohols; in fact, the molar ratio of alcohols to triglycerides is one of the most important parameters in the process. While greater proportions of alcohol favor more rapid and extensive reaction, they also create a greater volume of alkali waste to be treated; optimal ratios have therefore been reported ranging from 6:1 to 30:1, varying with the nature of the fat or oil and alcohols in use. The optimization of the alcohol : triglyceride molar ratio is also a key challenge for enzymatic catalysis [1, 7, 18].

Table 3. Physical and chemical properties of biodiesel in comparison to petroleum-derived diesel fuel. Adapted from [1]

Vegetable oil methyl ester	Kinematic viscosity (mm2/s)	Cetane number	Lower heating value (MJ/l)	Cloud point (°C)	Flash point (°C)	Density (g/l)	Sulfur (wt percent)
Peanut	4.9 (37.8°C)	54	33.6	5	176	0.883	--
Soybean	4.5 (37.8°C)	45	33.5	1	178	0.885	--
Soybean	4.0 (40°C)	45.7-56	32.7	--	--	0.880 (15°C)	--
Babassu	3.6 (37.8°C)	63	31.8	4	127	0.879	--
Palm	5.7 (37.8°C)	62	33.5	13	164	0.880	--
Palm	4.3-4.5 (4O°C)	64.3-70	32.4	--	--	0.872-0.877 (15°C)	--
Sunflower	4.6 (37.8°C)	49	33.5	1	183	0.860	--
Tallow	--	--	--	12	96	--	--
Rapeseed	4.2 (40°C)	51-59.7	32.8	--	--	0.882 (15°C)	--
Used rapeseed	9.48 (30°C)	53	36.7	--	192	0.895	0.002
Used corn oil	6.23 (30°C)	63.9	42.3	--	166	0.884	0.0013
Diesel fuel	12-3.5 (40°C)	51	35.5	--	--	0.830-0.840 (15°C)	--
JIS-2D (Gas oil)	2.8 (30°C)	58	42.7	--59		0.833	0.05

Transesterification can be accomplished at a variety of temperatures, with temperature optima depending on the oils involved; process temperatures of 25–100°C are common [7]. Catalysis of the process is essential (except in supercritical solvents) and may be accomplished by acids, bases, or enzymes: alkali catalysis employing NaOH, KOH, or a corresponding alkoxide is typically the most rapid and efficient in fats and oils with extremely low-water contents and low concentrations of free fatty acids [1, 7]. While this process is the one currently predominant in industry, it does have important limitations in the energy required to recover the glycerol by-product and in treating the alkali waste, as well as in the interference of water and free fatty acids with reaction progress [18].

If either water or free fatty acids are present in unacceptably high amounts, acid catalysis is preferable, typically employing hydrochloric, sulfuric, or phosphoric acids. While this process proceeds several thousand-fold more slowly than the alkali-catalyzed process, it does allow use of lower-quality oils [1].

2.2.4. Supercritical methanol catalysis. Recently, researchers have attempted transesterification in supercritical methanol, where supercritical fluids are solvent phases obtained under high temperature and pressure that have both vapor and liquid characteristics and in which many reactions have been found to proceed more readily than in subcritical solvents [19]. Since supercritical methanol has a hydrophobic nature with a lower dielectric constant than subcritical methanol, nonpolar triglycerides were well-solvated with supercritical methanol to form a single-phase oil-methanol mixture. As a result, the oil to methyl ester conversion rate increased dramatically in the supercritical state. Added advantages were that free fatty acids in the oil could also be converted efficiently to the methyl esters, improving product yield, and that purification of products following transesterification was much simpler and more environmentally friendly due to the absence of alkali (or acid) catalysts. Unfortunately, however, the process required a temperature of 3 50°C and pressure of 45MPa, as well as large quantities of methanol, with the result that much optimization will be required before this approach is commercially competitive [20–22]

2.3. Enzymatic (Lipase-based) Transesterification

In efforts to improve upon the abiotic transesterification methods, considerable effort has been devoted to the investigation of lipases, carboxylesterases that catalyze the hydrolysis and synthesis of long-chain acylglycerols, for the synthesis of biodiesel [23]. These remarkable enzymes currently constitute the single most widely-used biocatalyst class in biotechnology: they often possess high chemoselectivity, regioselectivity, and stereoselectivity; they are naturally extracellular enzymes and many are secreted in great quantity by fungi and bacteria, allowing relatively simple purification from culture media; they require no cofactors; they typically catalyze no undesirable side reactions; and the crystal structures of a number of representatives have been solved, facilitating rational design approaches in optimizing enzyme activity. The commercial lipase market is currently approximately 1 billion dollars per year, involving applications in detergents and in the production of food ingredients and enantiopure pharmaceuticals [24]; this status is fortunate for biodiesel applications because it provides preexisting incentives for improvements in lipase production and activity. A chart comparing the relative merits of the alkali-catalyzed and enzyme-catalyzed transesterification processes for biodiesel production is shown in Table 4.

Commercial lipases on which much biodiesel research has been based include enzymes from Mucor miehei (Lipozyme IM-20) and Candida antarctica (Novozym 435), as well as from Pseudomonas cepacia, Candida rugosa, and Rhizopus delemar. Both extracellular and intracellular lipases effectively catalyze a variety of triglyceride transesterification reactions, and they have the potential to overcome all of the problems of chemical catalysis (Table 4). On the other hand, the cost of lipase catalysts is significantly greater than that of alkali catalysts [1].

Table 4. Comparison of alkali- and enzymatically-catalyzed transesterification for biodiesel production. Adapted from [1]

Parameter	Alkali-catalyzed process	Lipase-catalyzed process
Reaction temperature	60-70°C	30-40°C
Free fatty acids in raw materials	Formation of undesired saponified products	Formation of desired methyl esters
Water in raw materials	Interference with the reaction	No influence
Yield of methyl esters	Normal	Higher
Recovery of glycerol	Difficult	Easy
Purification of methyl esters	Repeated washing required	No washing required
Production cost of catalyst	Low	Relatively high

Studies from the mid-1990s to present showed that, in general, lipase-catalyzed reactions with longer-chain fatty alcohols proceeded far more readily than those with methanol or ethanol. The reactions tolerated the presence of up to 20 percent water but were favored by the presence of organic solvents, which presented problems in that organic solvents were not suitable for fuel production due to the risk of explosion as well as the difficulty of removing the solvent. In addition, scientists found that even immobilized lipase preparations could not be re-used, a problem that would have to be solved to promote commercialization [18]. The greatest obstacle to this technology is the cost of the lipase enzymes. Two primary approaches are in progress to address this difficulty: first, genetic engineering of microorganisms to produce lipases in greater quantities and with greater activities, and second, the investigation of whole-cell systems to allow in situ regeneration of the catalytic units.

2.3.1. Lipase engineering: production. The cost of lipase production is the primary impediment to commercialization of lipase-catalyzed systems. This is a general problem of enzyme-catalyzed processes, which industrial biotechnology has addressed primarily by the development of high-enzyme-expression systems and of whole-cell biocatalysts.

Overexpression of lipases requires accurate protein folding and translocation across the cytoplasmic and, in gram-negative bacteria, outer membranes, processes estimated to involve up to 30 cytoplasmic proteins in some organisms. Through careful manipulation of lipase signal sequences and secretory pathways, however, several successes have been achieved: lipases from various Bacillus species have been overexpressed in E. coli systems, for example [24]; Rhizopus oryzae lipases have been produced in extracellular, functional form in Saccharomyces cerevisiae [25, 26]; and the Candida antarctica lipase B has been similarly secreted in functional form from the yeast Pichia pastoris [27]. Because lipase folding and secretion are highly specific processes that normally do not function properly in heterologous hosts, these successes represent breakthroughs that provide wonderful opportunities for further optimization and increases of lipase overexpression in heterologous hosts.

Work directed toward modifying and accelerating the secretion pathways in native lipase hosts has also led to great increases in extracellular lipase yield within Pseudomonas fluorescens and Serratia marcescens, showing that pathway manipulation within native hosts may be another promising avenue for further improvement [24].

2.3.2. Lipase engineering: activity. Another important challenge faced by lipase use in biodiesel production has been the diminished efficiency of lipases in transesterifications with methanol and ethanol in comparison to longer-chain alcohols. Lipases typically accomplish most rapid catalysis when the substrates are freely soluble in one another; methanol and ethanol, however, were found to be only soluble in vegetable (soybean and rapeseed) oils at molar ratios of 1:2 and 2:3 (alcohol : triglyceride fatty acid), respectively. Moreover, insoluble methanol caused rapid inactivation of the lipases. These realizations led to the development of a system in which alcohols were added in discrete doses, below their solubility limits, to batch reaction mixtures, improving the reaction yield to near-completion (>97 percent) as well as extending the lifetime of the lipases to greater than 100 days [1, 18].

Improving the tolerance of lipases to methanol or ethanol is also possible: certain species of Fusarium, Pseudomonas, and Bacillus produce solvent-tolerant lipases [1], and directed evolution has the potential to effect further desired changes in solvent tolerance, catalytic rate, and substrate specificity in these or other enzymes [24], reviewed in [28]. Immobilization of lipases within gels may extend lipase activity, as well; work of Noureddini and colleagues in screening a number of extracellular lipases for methanoloysis and ethanolysis of soybean oil revealed that several of the enzymes were more active, and retained their activities for longer periods of time, when subjected to such immobilization [29], building on previous work investigating lipase immobilization for transesterifiction [30, 31].

2.3.3. Whole-cell systems. To diminish the time and energy requirements of enzyme purification, even of a relatively simple purification such as that required with lipases, as well as problems presented by instability of extracellular enzymes, significant effort has addressed the direct use of whole cells as biocatalysts [32–34]. For example, lipases can be overexpressed within biotechnologically tractable hosts in which they are not secreted, followed by permeabilization of the host to allow catalysis to occur within the (compromised) cytoplasm. Cytoplasmic overexpression of the Rhizopus oryzae lipase in Saccharomyces cerevisiae, followed by freeze-thawing and air drying of the yeast cells, resulted in a whole-cell biocatalyst that effectively catalyzed triglyceride methanolysis [35]; optimization of the membrane fatty acid composition further improved lipase activity and stability [36].

In addition, cells producing their native lipases have been recruited. In work directed toward whole-cell lipase optimization by Ban and colleagues, cells were immobilized using porous biomass support particles (BSPs) made of polyurethane foam by introducing the BSPs during batch cultivation and allowing the cells to colonize them spontaneously. Cells were then cross-linked onto the support with glutaraldehyde, a procedure that greatly extended the enzymatic activity longevity. Once immobilized in this form, the particles could be treated much as conventional solid-phase catalysts: aseptic handling of the particles was unnecessary, the particles could withstand mechanical shear, and they could be reused for up to 6 batches.

Mass transfer rates were also sufficiently rapid within the BSPs that conversion rates approached those obtained with extracellular lipases: methanolysis of soybean oil by immobilized Rhizopus oryzae cells, in the presence of 10–20 percent water, reached 80–90 percent without any organic solvent pretreatment [37, 1, 38]; additional improvements in catalytic rate and durability were reported recently with this system in an air-lift reactor configuration [39]. Further development of whole-cell biocatalysts is thus positioned to make important contributions to biodiesel production.

3. Research Priorities

The cost, quality, and performance of biodiesel, as well as its overall environmental profile, could be improved by further efforts in several areas. First, feedstocks other than virgin plant oils, most of which are cultivated by non-sustainable, pesticide and energy intensive agricultural practices, would ideally be explored and developed; alternatively, sustainable cultivation of oil crops should be developed. Waste oil processing technology also deserves developmental effort to allow recovery of its intrinsic energy, and microbial and algal lipid production should be investigated to determine whether they might provide feedstocks at lower cost.

Second, the further development of lipase technology will facilitate efficient enzymatic transesterifications of feedstock oils and fats and production of benign wastes with easily-recoverable coproducts, principally glycerol. Specifically, genetic engineering of lipases for greater activity and durability, as well as metabolic engineering of lipase-production pathways to understand lipase synthesis and regulation and to facilitate extracellular production, are well-positioned to offer valuable advances in enzymatic transesterification of oils and should be pursued. Research in these areas could have potentially great impacts within relatively short timescales and should be encouraged to the greatest extent possible.

4. Commercialization

International commercialization of biodiesel is well underway in both the United States and in many European and Asian countries. Recent news announcements by the U.S. National Biodiesel Board, revealing the growing trend in adoption of biodiesel fuels even in this country, include the following: the announcement on May 28, 2004, by World Energy Alternatives LLC of the re-opening and upgrading of the largest multi-feedstock biodiesel production facility in the United States, with a capacity of 18 million gallons, in Florida; the opening of Canada's first retail biodiesel pump in Toronto by Topia Energy, Inc. on March 2, 2004; the opening of 10 biodiesel pumps in Denver, Colorado in May 2004 by Blue Sun Biodiesel as part of a citywide pilot program; and the adoption of biodiesel for its maintenance vehicles by the Big South Fork National River and Recreation Area in Tennessee, joining dozens of other U.S. national parks [40].

Biodiesel is most readily available in B20, B50, and B100 blends, representing 20, 50, and 100 percent biodiesel mixed with a balance of petroleum diesel, respectively. These fuels are widely available in the United States, particularly in coastal areas and in midwest agricultural regions (a current map of U.S. retail biodiesel locations is provided at http://biodiesel.org/ buyingbiodiesel/retailfuelingsites/default.shtm/) and can be used in conventional diesel engines such as those found in the Volkswagen Golf, Jetta, Jetta Wagon, New Beetle, and 2004 Passat; 2004 Mercedes E-320 Sedan; 2004 Chrysler Jeep Liberty and 2004 Volkswagen Touareg SUV; Chevy Silverado; GMC Sierra; Dodge Ram; and Ford E-series and F-series trucks.

Biodiesel production capacity exists and is expanding, and technology for storage and usage is in place. The primary deterrent to even more widespread commercialization is simply its price, which is increasingly comparable to petroleum diesel as oil prices rise and tax incentives are enacted.

In contrast to other biofuels, biodiesel is already an established product with an established set of technologies (both for synthesis and for consumer use) supporting it. Because of this, further improvements in the cost-effectiveness of its production, in combination with rising emissions standards, will have significant impacts on its attractiveness to consumers.

REFERENCES

[1]   Fukuda, H., A. Kondo, and H. Noda (2001). *Biodiesel fuel production by transesterification of oils,* J Biosci Bioeng 92:405-416.

[2]   Pearl, G. G. (2001). Biodiesel Production in the US, Render Magazine, http://www.rendermagazine.com/August2001/TechTopics.html.

[3]   Dmirbas, A. (2003). *Biodiesel fuels from vegetable oils via catalytic and non-catalytic supercritical alcohol transesterifications and other methods: A survey,* Energy Conv Mgmt 44:2093-2 109.

[4]   National Biodiesel Board (2004). Biodiesel FAQs, National Biodiesel Board, http://www.biodiesel.org/resources/faqs/.

[5]   Chevron Corporation (2004). Dictionary of Lubricant Terms, http://www.chevron.com/ oronite/reference_materials/dictionary _of_lubricant_terms/li_dictionary _d.asp.

[6]   Wikipedia (2004). Cetane, Wikipedia, the Free Encylopedia, http://www.riograndesoftware.com/encyclopedia/c/ce/cetane.html.

[7]   Ma, F., and M. A. Hanna (1999). *Biodiesel production: A review,* Bioresource Technol 70: 1-15.

[8]   Allen, C. A. W., K. C. Watts, R. G. Ackman, and M. J. Pegg (1999). *Predicting the viscosity of biodiesel fuels from their fatty acid ester composition,* Fuel 78:1319-1326.

[9]   TA Travel Centers (2004). Diesel Fuel Prices, http://www.tatravelcenters.com/ta/ display _diesel _fuel_prices.phtml.

[10]   Chicago Board of Trade (2004). Real-time quotes, charts, and news, http://www.unitedsoybean.org/.

[11]   TA Travel Centers (2005). *Diesel Fuel Prices,* http://www.tatravelcenters.com/ta/ display _diesel _fuel_prices.phtml.

[12]   Chicago Board of Trade (2005). *Historical Data: Soybean Oil,* www.cbot.com.

[13]   See-Search Engines (2004). Fuel/diesel/petrol prices across Europe, http://www.see-search.com/business/fuelandpetrolpriceseurope.htm.

[14]   National Biodiesel Board (2005). Tax Incentive Fact Sheet, http://www.biodiesel.org/members/membersonly/files/pdf/fedreg/20041022_Tax_Incen ti ve _Fact _Sheet.pdf.

[15]   Raneses, A. R., L. K. Glaser, J. M. Price, and J. A. Duffield (1999). *Potential biodiesel markets and their economic effects on the agricultural sector of the United States,* Indust Crops Prod 9:151-162.

[16]   de Guzman, D. (2001). Biodiesel market strengthens as alternative to diesel fuel, http://www.worldenergy.net/pdfs/news stories/chemical _marker.pdf.

[17] Zhang, M. A., M. A. Dube, D. D. McLean, and M. Kates (2003). *Biodiesel production from waste cooking oil: 1. Process design and technological assessment,* Biore source Technol 89:1-16.

[18] Shimada, Y., Y. Watanabe, A. Sugihara, and Y. Tominaga (2002). Enzymatic alcoholysis for biodiesel fuel production and application of the reaction to oil processing, J Molec Catalysis B: Enzymatic 17:133-142.

[19] Rayner, C. (2004). What are supercritical fluids?, University of Leeds, http://www.chem.leeds.ac.uk/People/CMR/whatarescf.html.

[20] Kusdiana, D., and S. Saka (2001). *Kinetics of transesterification in rapeseed oil to biodiesel fuel as treated in supercritical methanol,* Fuel 80:693-698.

[21] Kusdiana, D., and S. Saka (2001). *Methyl esterification of free fatty acids of rapeseed oil as treated in supercritical methanol,* J Chem Eng Jpn 34:383-3 87.

[22] Saka, S., and D. Kusdiana (2001). *Biodiesel fuel from rapeseed oil as prepared in supercritical methanol,* Fuel 80:225-231.

[23] King, M. W. (2003). *Fatty Acid Oxidation,* Indiana University School of Medicine, http://www.med.unibs.it/~marchesi/fatox.html.

[24] Jaeger, K.-E., and T. Eggert (2002). *Lipases for biotechnology,* Curr Opin Biotechnol 13 :390-397.

[25] Takahashi, S., M. Ueda, H. Atomi, H. D. Beer, U. T. Bornscheuer, R. D. Schmid, and A. Tanaka (1998). *Extracellular production of active Rhizopus oryzae lipase by Saccharomyces cerevisiae,* J Ferment Bioeng 86:164-168.

[26] Ueda, M., S. Takahashi, M. Washida, S. Shiraga, and A. Tanaka (2002). *Expression of Rhizopus oryzae lipase gene in Saccharomyces cerevisiae,* J Molec Catal B: Enz 17:113 - 124.

[27] Rotticci-Mulder, J. C., M. Gustavsson, M. Holmquist, K. Hult, and M. Martinelle (2001). *Expression in Pichia pastoris of Candida antarctica lipase B and lipase B fused to a cellulose-binding domain,* Protein Expression and Purification 21:386-392.

[28] Powell, K. A., S. W. Ramer, S. B. del Cardayre, W. P. C. Stemmer, M. B. Tobin, P. F. Longchamp, and G. W. Huisman (2001). Directed evolution and biocatalysis, Angew Chem Int Ed Engl 40:3948-3959.

[29] Noureddini, H., X. Gao and R.S. Philkana (2005). *Immobilized Pseudomonas cepacia lipase for biodiesel fuel production from soybean oil,* Bioresource Technol 96:769-777.

[30] Samukawa, T., M. Kaieda, T. Matsumoto, K. Ban, A. Kondo, Y. Shimada, H. Noda, and H. Fukuda (2000). *Pretreatment of immobilized Candida antarctica lipase for biodiesel fuel production from plant oil,* J Biosci Bioeng 90:180-183.

[31] Iso, M., B. Chen, M. Eguchi, T. Kudo, and S. Shrestha (2001). *Production of biodiesel fuel from triglycerides and alcohol using immobilized lipase,* J Molec Catal B: Enzym 16:53-58.

[32] Liu, Y., H. Hama, Y. Fujita, A. Kondo, Y. Inoue, A. Kimura, and H. Fukuda (1999). *Production of S-lactoyl-glutathione by high activity whole cell biocatalysts prepared by permeabilization of recombinant Saccharomyces cerevisiae with alcohols,* Biotechnol Bioeng 64:54-60.

[33] Kondo, A., Y. Liu, M. Furuta, Y. Fujita, T. Matsumoto, and H. Fukuda (2000). *Preparation of high activity whole cell biocatalyst by permeabilization of recombinant flocculent yeast with alcohol,* Enzyme Microb Technol 27:806-811.

[34] Liu, Y., Y. Fujita, A. Kondo, and H. Fukuda (2000). *Preparation of high-activity whole cell biocatalysts by permeabilization of recombinant yeasts with alcohol*, J Biosci Bioeng 89:554-558.

[35] Matsumoto, T., S. Takahashi, M. Kaieda, M. Ueda, A. Tanaka, H. Fukuda, and A. Kondo (2001). *Yeast whole-cell biocatalyst constructed by intracellular overproduction of Rhizopus oryzae lipase is applicable to biodiesel fuel production*, Appl Microbiol Biotechnol 57:515-520.

[36] Hama, S., H. Yamaji, M. Kaieda, M. Oda, A. Kondo, and H. Fukuda (2004). *Effect of fatty acid membrane composition on whole-cell biocatalysts for biodiesel-fuel production*, Biochem Eng J 21:155-160.

[37] Ban, K., M. Kaieda, T. Matsumoto, A. Kondo, and H. Fukuda (2001). *Whole cell biocatalyst for biodiesel fuel production utilizing Rhizopus oryzae cells immobilized within biomass support particles*, Biochem Eng J 8:39-43.

[38] Ban, K., S. Hama, K. Nishizuka, M. Kaieda, T. Matsumoto, A. Kondo, H. Noda, and H. Fukuda (2002). *Repeated use of whole-cell biocatalysts immobilized within biomass support particles for biodiesel fuel production*, J Molec Catal B: Enz 17:157-165.

[39] Oda, M., M. Kaieda, H. Hama, H. Yamaji, A. Kondo, E. Izumoto, and H. Fukuda (2005). *Facilitatory effect of immobilized lipase-producing Rhizopus oryzae cells on acyl migration in biodiesel-fuel production*, Biochem Eng J 23 :45-51.

[40] National Biodiesel Board (2004). In the News, National Biodiesel Board, http://www.biodiesel.org/.

[41] United States Department of Energy (2004). *Diesel Fuel Ignition*, Office of Energy Efficiency and Renewable Energy, http://www.ott.doe.gov/images/dieselpiston.jpg.

C. BIOHYDROGEN

1. Introduction

Molecular hydrogen (H_2) is a promising future energy source due to its clean combustion and to its potential for sustainable production [1, 2]. While the challenges in converting commercial processes based on hydrocarbon fuels to ones powered by hydrogen fuel cells are great, including major modifications in fuel storage and transport infrastructure, the potential advantages of hydrogen are sufficiently great to have drawn extensive attention to the research and development of H_2 production and utilization technologies [3, 4].

Currently, H_2 is produced primarily by electrolysis of water, requiring a source of electricity, and steam reforming of natural gas, requiring both a nonrenewable fossil fuel feedstock and additional energy to create the necessary heat and pressure [2, 5]. To improve the environmental profile of H_2 production, numerous other technologies are being developed as well. Among these, microbial mechanisms that obtain energy either through photosynthesis (via photosynthetic, nitrogenase-mediated, or photo-fermentative pathways) or through consumption of organic substrates, potentially including organic wastes (water-gas shift and dark fermentative pathways), are of particular interest because of their potentially low requirements for expensive and non-renewable energy sources [6].

2. State of the Science

2.1. Microorganisms

Several types of microorganisms, undergoing different, yet related metabolisms, are involved in H2 production. To aid in clarifying and distinguishing these mechanisms, let us first consider the microorganisms involved.

2.1.1. Oxygenic phototrophs. Oxygenic phototrophs include both aerobic eukaryotes (plants and green algae) and aerobic prokaryotes (cyanobacteria) that are able to use photons to energize electrons from water, thus yielding O_2, for the ultimate purpose of generating the reduced form of nicotinamide adenine dinucleotide phosphate (NADPH) for CO_2 fixation into organic carbon (Figure 16). The electrons are energized by passage through two successive photosystems, photosystems II (PS II) and I (PS I), each of which harvests photons with networks of chlorophyll and other accessory pigments and uses the photon energy to impart greater reducing power to the electrons. After nicotinamide adenine dinucleotide phosphate (NADP) is reduced to NADPH, and CO_2 is fixed into organic carbon, phototrophs typically respire this substrate aerobically to generate adenosine triphosphate (ATP).

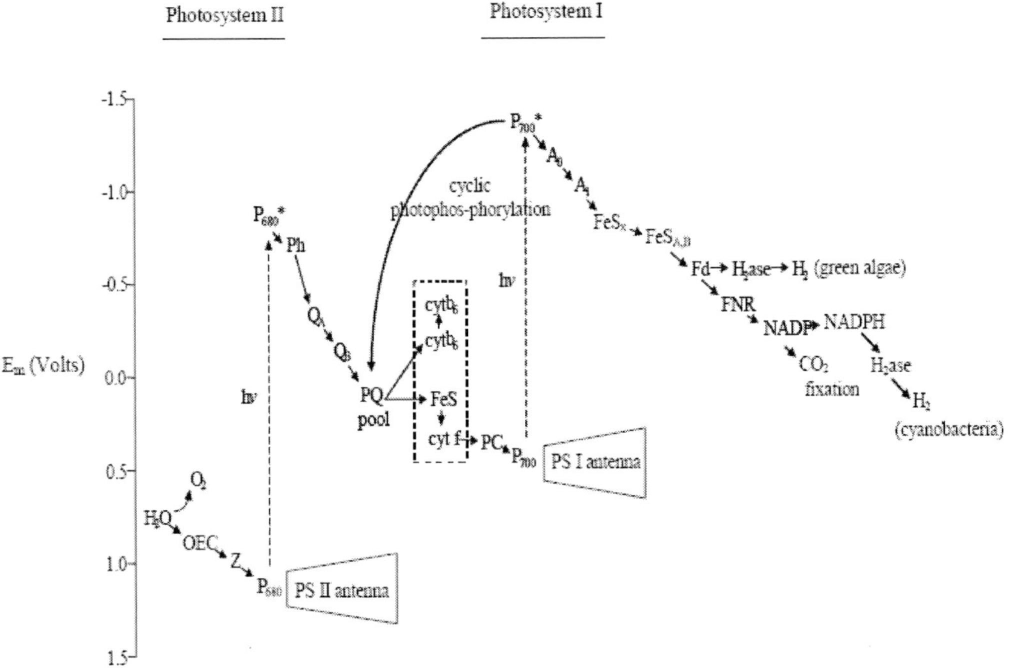

Dashed arrows represent light reactions that extract electrons from H_2O and deliver them to ferredoxin (Fd), from which they can be used to fix CO_2 via NADPH or to produce H_2 via hydrogenases. Abbreviations represent common electron carriers, arranged to show their relative redox potentials, Em.

Figure 16. The Z-scheme for electron transport in the oxygenic photosynthesis of cyanobacteria and green algae.

Alternatively, electrons may also be returned by photosystem I to the plastoquinone pool in a process called cyclic photophosphorylation (Figure 16). This process allows additional protons to be pumped through the thylakoid membrane, contributing to the chemiosmotic gradient that drives ATP production [7].

2.1.1.1. Green algae. Green algae are unicellular eukaryotes in which the majority of H_2-producing capability lies in the use of photosynthetically-activated electrons by Fe-hydrogenase enzymes, requiring anaerobiosis in light. This process is known as direct photolysis. Green algae typically also have significant H2 uptake activity, although the enzymes responsible have not yet been identified. In general terms, therefore, the primary enzymes of concern for H_2 production in green algae are Fe-hydrogenases, known to produce H2, and uptake hydrogenases of unknown composition.

Green algae, as well as cyanobacteria (below), must withstand periods of anaerobiosis during darkness and therefore are also capable of fermentation (Figure 17). While reducing power generated by fermentation of endogenous substrates can enter the plastoquinone pool, undergo activation by PS I, and lead to H2 production, H2 generation through fermentation is estimated to have only 1/100 of the potential of direct photolysis in green algae [9, 10]. Nevertheless, the presence of this fermentative metabolism is physiologically significant for H_2 production by direct photolysis, as it appears that the ability of these organisms to evolve H_2 by means of hydrogenases originated with the necessity of disposing of excess reducing power during fermentation [11].

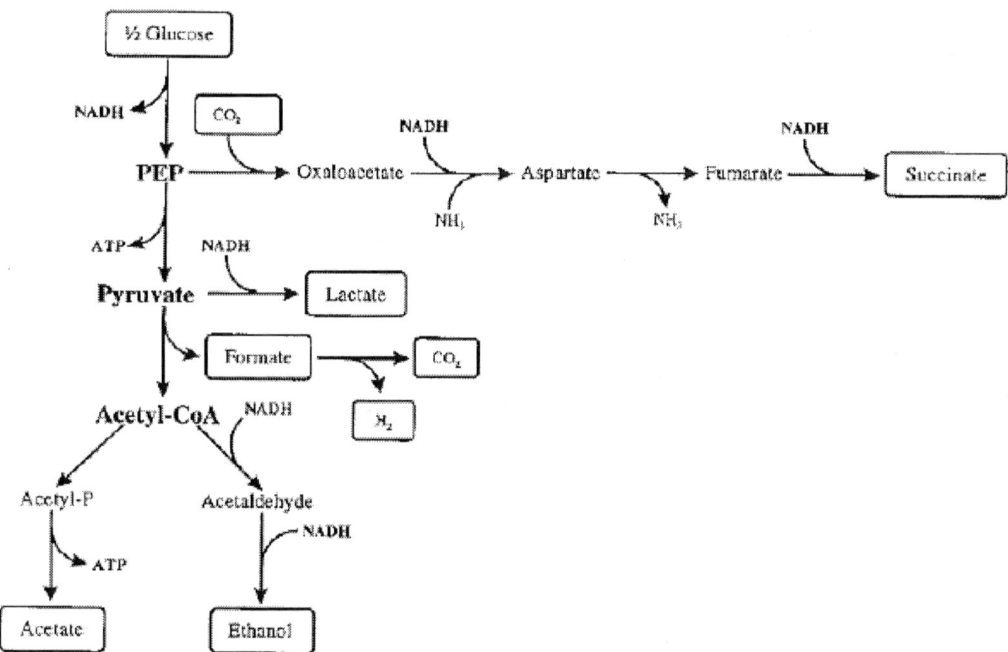

Figure 17. The mixed-acid fermentation of Escherichia coli, an example of a common H_2-producing fermentative pathway, showing H2 production from formate [8].

2.1.1.2. Cyanobacteria. Cyanobacteria or blue-green algae, in contrast to green algae, are unicellular or multi-cellular (filamentous) prokaryotes that generate H_2 by three distinct mechanisms. These pathways can, under certain conditions, operate simultaneously. Among N2-fixing cyanobacteria, the majority of H_2-producing capability lies in the activity of nitrogenase enzymes that use photosynthetically-generated ATP to reduce or "fix" molecular nitrogen and simultaneously to generate H_2 [12]; this process is known as indirect photolysis (Figure 18).

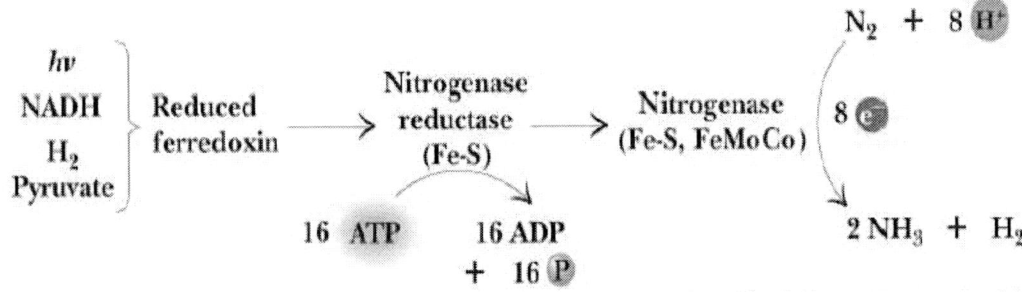

Figure 18. Nitrogenase reaction equation, showing electron donors, the role of ATP, and H2 production from reduced protons [16].

The cyanobacterium Synechocystis sp. strain PCC 6803 M55, lacking genetic evidence of both nitrogenase and uptake hydrogenases, has been found recently to exhibit direct photolysis [13], opening the possibility that direct photolysis may be found in other cyanobacteria as well. Cyanobacteria can also metabolize endogenous substrate anaerobically through a number of fermentation pathways (e.g., Figure 17), several of which produce H2 by means of bidirectional hydrogenases [14, 15].

2.1.2. Anoxygenic phototrophs. Anoxygenic phototrophs, in contrast to oxygenic phototrophs, are exclusively prokaryotic and often obligately anaerobic. These microbes each possess only one photosystem: some bacterial reaction centers are analogous to PS I, and some are analogous to PS II, but none extract electrons from water, and thus O_2 is not produced (Figure 19). Instead of water, the reaction centers must energize electrons from organic or inorganic substrates found in their environments. Although many photosynthetic bacteria depend on Rubisco and the Calvin cycle for the reduction of CO_2, some are able to fix atmospheric CO_2 by other biochemical pathways. Despite these differences, energy transduction is carried out by mechanisms quite similar to those found in oxygenic phototrophs: the light- harvesting centers contain bacteriochlorophylls, analogous to the chlorophylls but with strongest absorption in the infrared (700–1000 nm), as well as carotenoids, and electron transport proteins are also analogous. As in oxygenic photosynthesis, electron transfer is coupled to the generation of an electrochemical potential that drives phosphorylation by ATP synthase, and the energy required for the reduction of CO2 is provided by ATP and NADH, a molecule similar to NADPH [17].

Anoxygenic phototrophs fall into several categories, based on pigmentation and electron donor. Of greatest interest to biohydrogen production are the purple non-sulfur bacteria, e.g., members of the genera Rhodobacter and Rhodopseudomonas, which use organic compounds or H_2 as electron donors [17]. These microbes prefer to grow photoheterotrophically, using simple fatty acids such as succinate and malate to supply electrons for photosynthesis as well as carbon for biosynthesis. In these organisms, most of which are able to fix nitrogen [18], H2 production is associated primarily with nitrogenase activity [19].

2.1.3. Fermenters. Fermentations are energy-yielding metabolic processes in which an organic substrate is decomposed into smaller molecules, some of which are oxidized and some of which are reduced relative to the original substrate.

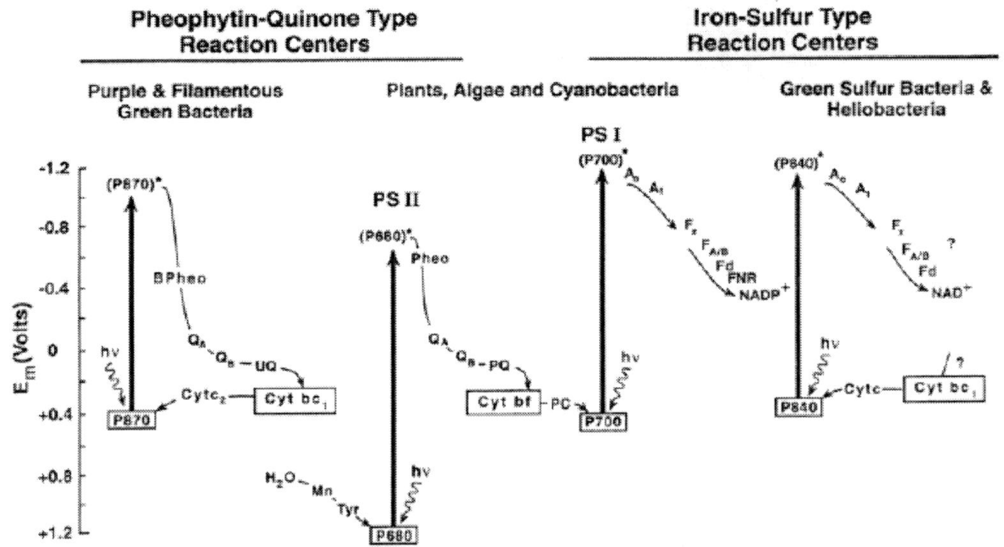

Figure 19. Relationships among photosystems of oxygenic (plants, algae, and cyanobacteria) and anoxygenic (purple bacteria, green filamentous bacteria, green sulfur bacteria, and heliobacteria) [17].

Fermentations occur in the absence of an electron transport chain or exogenous terminal electron acceptors such as O_2 [8]. Fermentations are therefore anaerobic processes, but they may be carried out by either anaerobic or aerobic microorganisms within sufficiently anoxic microenvironments [20]. Numerous fermentation pathways are known, many of which, like the mixed acid fermentation pathway (Figure 17), produce H_2, and all of which generate multiple products that typically must be separated and purified for commercial use [8, 14, 5]. While members of both the oxygenic and anoxygenic phototrophs possess H_2-producing fermentation pathways, only bacteria appear to produce H_2 from fermentation in appreciable quantities. These bacteria include members of the Bacillus, Clostridia, and Enterobacter, as well as the extreme thermophiles Caldicellulosiruptor and Thermotoga, all heterotrophs that thrive on carbohydrates and typically are not capable of anaerobic respirations [5].

2.2. Enzymes

Two categories of enzymes, the hydrogenases and the nitrogenases, are responsible for H_2 production in the microorganisms discussed above.

2.2.1. Hydrogenases. Hydrogenases catalyze the redox interconversion of protons with H_2 gas:

$$2H^+ + 2e^- \Leftrightarrow H_2$$

They occur in two primary forms, known as uptake and bidirectional hydrogenases. Uptake hydrogenases oxidize H_2 to H+ to provide reducing power to many anaerobic respiratory organisms [21], while bidirectional hydrogenases may catalyze either H_2 oxidation or H+ reduction, although in vivo they typically function in only one of the two capacities [22, 23]. Hydrogenases may also be classified according to differences in their organometallic catalytic sites: NiFe(Se)-hydrogenases are characterized by catalytic sites possessing coordinated nickel, iron, sulfur, and in some cases selenium [21], while Fe-hydrogenases contain only Fe-S centers in their active sites [23]. Metal-free hydrogenases also exist [24]. Many of the known bidirectional hydrogenases are Fe-hydrogenases [23], while the majority of uptake hydrogenases are NiFe-hydrogenases; Fe-hydrogenases functioning to produce H2 are thus of greatest interest in microbial H2 production. Nevertheless, uptake hydrogenases are of equal or perhaps greater concern in some systems, because they frequently co-exist with H_2-producing hydrogenases and can recycle H2 within the microbe, greatly diminishing overall H_2 yield [12].

The physiological role of hydrogenase-based H2 production appears to be the discharge of excess reducing power, necessary when other suitable electron acceptors such as O2 are absent [25-27]. It is not surprising, therefore, that Fe-hydrogenases are rapidly inhibited both transcriptionally and post-translationally by molecular oxygen [28, 23]. This feature has significant implications for biological H_2 production, requiring anoxic cellular environments to be maintained for both induction of hydrogenase synthesis as well as for continued hydrogenase activity. For fermentative H_2 production, this is no obstacle, as the entire microbial metabolism takes place anaerobically. However, oxygenic phototrophs provide the majority of activated electrons to Fe-hydrogenases through the O_2-generating photosynthetic process [29]. Because Fe-hydrogenases are inhibited both transcriptionally and post-translationally by O_2, photosynthetic electron production and hydrogenase-based H_2 production cannot occur simultaneously in a wild-type organism. These processes therefore must be separated either temporally or spatially, and much research in biohydrogen production is directed toward the accomplishment of these goals.

Fe-hydrogenases exist in monomeric, dimeric, and at least one trimeric form [30, 23]. The two C. *reinhardtii* Fe-hydrogenase enzymes cloned and sequenced to date [31] encode enzymes that are among the smallest hydrogenases known. Consistent with other algal hydrogenases, they contain only the single catalytic Fe-S center, or H-cluster, and ferredoxin is the only putative electron donor. In contrast, bacterial hydrogenases typically contain several additional iron-sulfur centers and accept electrons from a variety of donors. Clostridial hydrogenases, involved in hydrogen evolution during the fermentation of carbohydrates, can accept electrons from flavodoxins, for example [24].

2.2.2. Nitrogenases. Nitrogenases are tetrameric organometallic enzymes that catalyze reductive cleavage of the second-strongest chemical bond known:the triple bond of dinitrogen gas, N_2. To supply the tremendous energy needed for this process, the enzyme uses ATP; additional activated electrons are also required, however, which may be delivered through NADH, pyruvate, photoactivation, or H2 itself (Figure 18). Nitrogenases share many characteristics with hydrogenases: they employ metallic catalytic centers to facilitate the redox reaction (Fe-Mo-Co and Fe-S clusters); they are rapidly inactivated by exposure to molecular oxygen; and their synthesis is tightly regulated, requiring N deficiency as well as anaerobiosis. Nitrogenases are found in many cyanobacteria, most of the purple non-sulfur bacteria, and numerous symbiotic and free-living eubacteria. They evolve H2 rapidly during active nitrogen fixation; however, this H2 is produced at the cost of 16 ATP per H2 [32]. Nitrogenases appear to produce the majority of H2 in nitrogen-fixing cyanobacteria (indirect photolysis) and purple non- sulfur bacteria (photofermentation) [19].

2.3. Direct Photolysis

Direct photolysis is the process in which bidirectional hydrogenases use photosynthetically-activated electrons, via reduced Fd and/or NADPH, to reduce the hydrogen ion (H+)to H_2 [28] (Figure 16). This process is carried out by green algae such as Chlamydomonas reinhardtii, Scenedesmus, and Chlorella [22, 33, 28], as well as by cyanobacteria such as Synechocystis [13]. Light-dependent H2 production by cyanobacteria utilizing nitrogenases and photosynthetically-generated ATP is known as indirect photolysis, and light-dependent H2 production by anoxygenic phototrophs utilizing organic electron donors, also making use of nitrogenases, is known instead as photofermentation.

2.3.1. Photosynthetic efficiency. Energy efficiency, defined as the ratio of energy produced as H2 to the resources consumed by the microorganism (including its requirements for space as well as nutrients), is a central consideration guiding research directions and assessment of commercial applicability of biohydrogen systems. The rate of H2 production is an equally important concern, however, and systems of lower efficiency but higher H2 production rate may be competitive with those of higher efficiency in cases where substrate and space costs are low [34, 5].

For photochemical processes, Einstein's law of photochemistry states that a primary photochemical process is caused by the action of one absorbed photon acting on another molecule, emphasizing the important fact that a photon must first be absorbed before it can carry out photochemistry. The next important parameter is the photochemical quantum yield, Φ, defined as the ratio of the number of photochemical products to the number of absorbed protons. The quantum yield is a measure of the efficiency of the photochemical proces: the primary quantum yield ranges from one for a process in which every absorbed photon leads to products, to zero when no products are formed. In photosynthetic systems, the primary quantum yields are often close to one under optimal conditions, indicating that almost all absorbed photons are effective in forming initial products. However, more than one photon is usually needed to produce a stable final product such as H_2, so overall quantum yields are usually much less than one [7].

As a consequence in part of the high primary quantum yield, direct photolysis has a very high potential energetic efficiency compared to other known systems [35–39]. This efficiency is estimated to be as high as 10 percent for microalgal photosynthesis resulting in CO_2 fixation [34] and as high as 24 percent for H2 production, under ideal conditions [35]. For comparison, a typical commercial steam turbine generator is about 30 percent efficient, photovoltaic cells also approach 30 percent efficiency, automobiles are ~20 percent efficient, and a well-tuned bicycle rates about 75 percent [40].

The important limitation to photosynthetic conversion efficiency lies in the difficulty of providing so-called ideal conditions to photosynthetic cultures, resulting in reported conversion efficiencies that are usually less than 1 percent [34].

2.3.2. Light saturation effect. A salient difficulty in providing ideal conditions for phototroph culture involves provision of the light itself. The microbial aspect of this problem lies in the ability of both algae and cyanobacteria to absorb many more photons with the light-harvesting antennae, and therefore to generate many more excited electrons, than the photosynthetic electron transport chain can accommodate. Although these microbes have evolved sensitive regulatory systems to optimize the size of their photon-gathering antennae (chlorophylls a and b, xanthophylls, phycobilins, etc.), diminishing them considerably under high light and increasing them under low light, phototrophs nevertheless typically absorb many more photons than they are able to use productively. Wild-type algae exposed to full sunlight absorb up to nine times more photons than they can accommodate, wasting the remainder in the form of heat or fluorescence and shading cells below them quite effectively. This phenomenon is known as the light-saturation effect and represents one of the prominent challenges facing commercialization of direct photolysis [34].

One promising approach to this problem is the creation of algal mutants with diminished light-harvesting antennae, with the logic that a photosynthetic apparatus with less light-harvesting capacity absorbs fewer photons at high light intensities and therefore wastes fewer photons. Japanese researchers working with microalgal mutants with reduced antenna sizes found increases in productivity of up to 50 percent under high light intensity, compared to the wild-type [41, 42]; antenna mutants isolated in the United States also showed efficiency improvements, further supporting the potential of this approach [43, 44]. Extensive cost analysis by modelers at the NREL confirmed the value of increased light transmission: increasing incident light transmission by a factor of 10, thought to be well within the capability of modern genetic techniques, reduced the cost of H2 production 57 percent [45].

2.3.3. Land use. The land-use aspect of the light provision problem is the relatively low density of solar energy, estimated at a maximum of 5 kilowatt hours per square meter per day or 6.6 gigajoules per square meter per year in the most favorable locations [34, 46]. Assuming a conversion of 10 percent of the solar energy into H2 and a price for H2 of $15 per gigajoule (the benchmark price set by DOE for H_2), this computes to ~0.66 gigajoules or ~$10 H_2 per square meter per year [34]. For comparison, an average electrically-heated house in North America consumes approximately 55–70 gigajoules of energy per year [5, 47, 48]. While this discrepancy seems large, it is worth noting that energy-efficient construction and practices, including higher-density housing, can reduce residential energy requirements substantially. Even without such measures, it is instructive to realize that, given a 10 percent solar conversion efficiency, the annual energy needs of the United States could be met by a square ~100 miles on a side located in a virtually unoccupied area of southern Nevada [49].

To address the problem presented by the limits of solar irradiance, potentially requiring large land areas for photobioreactors, efforts have been made to optimize vertical photobioreactor arrays [34], to investigate thin-layer bioreactors (0.5-5 centimeters in depth) with corresponding low masses that could be installed on rooftops [45], and most futuristically, to design solar collectors to transmit solar energy to bioreactors through optical fibers [50].

2.3.4. Oxygen sensitivity. The second primary difficulty in providing ideal conditions for direct photolysis lies in the extreme sensitivity of hydrogenases to the molecular oxygen produced by photosynthesis. As a result, cultures maintained under dark, anoxic conditions to induce hydrogenase synthesis are able to sustain H_2 production for only a few minutes following exposure to light and consequent photosynthetic O2 generation [51]. In response to this problem, two primary approaches are being taken: the first is the development of specialized bioreactors to separate O_2 generation from H2 generation temporally, and the second is the effort to generate O_2-tolerant hydrogenases through various mutagenic methods.

The most successful approach to date, to minimize the problem of O_2 inhibition of the algal Fe *hydrogenase,* has been the development of the two-stage photobioreactor. In this system, algal cultures experience two alternating growth conditions: the first supports photoautotrophic growth, in the presence of all essential nutrients as well as O_2, while the second deprives the cultures of both O2 and sulfur. Reduced sulfur is essential to synthesis of cysteine and methionine and is therefore essential to the synthesis of proteins as well, including the rapidly-recycled D1 protein of PS II. In the absence of reduced sulfur, PS II function diminishes significantly, resulting in decline of O2 production as well, even under light exposure. Algal aerobic respiration is not strongly affected by the sulfur deprivation over relatively short time periods (<100 hours) and therefore proceeds, diminishing O_2 concentrations sufficiently to induce the Fe-hydrogenase activity. Because PS I and the electron-transport proteins, cytochromes b6 and f, are not significantly affected, the cell can maintain an ATP-generating proton gradient across the thylakoid membrane by using PS I to activate electrons released to the plastoquinone pool (Figure 16) by the degradation of starch, proteins, and lipids via a NADH reductase complex. H2 production then follows, as a means of removing the spent electrons from the photosynthetic electron transport chain [11], for approximately 100h [52].

Commercialization of this process is being actively investigated [44]. Significant effort has also been directed toward the modification of Fe-hydrogenases to yield varieties with greater O_2 tolerance. This work has been encouraged by the discoveries of greater O_2 tolerance among some hydrogenases, particularly those of the bacterium Desulfovibrio vulgaris and the alga Pandorina morum [37], as well as the discovery of an Fe-hydrogenase sequence in the genome of the bacterium Shewanella oneidensis, a facultative rather than strict anaerobe (53). In addition, the solution of the crystal structures of both monomeric and dimeric Fe-hydrogenases [54, 55]. Flynn, Ghirardi, and colleagues employed traditional chemical mutagenesis and screening with H2-sensitive tungsten oxide films to isolate mutants with up to 10 times greater O_2-tolerance of H2 production [51, 29, 56–59] as well as increased rates of H2 production [60]. This success has inspired further efforts to alter one of the Chlamydomonas Fe-hydrogenases to diminish substantially its sensitivity to O_2. Toward this end, further random chemical mutagenesis, error-prone PCR-mediated mutagenesis, and site- directed mutagenesis are currently underway [61].

At least two other research groups are also investigating directed evolution as an approach for generating greater O_2 tolerance in algal Fe-hydrogenases. Because gene shuffling requires a diverse pool of parental Fe-hydrogenase genes, it is fortunate that numerous potential parental genes exist, representing genera among the archaea, eubacteria, fungi, algae, protists, and higher eukaryotes as well as monomeric and dimeric forms. Even among the multimeric enzymes, genes for subunits that show significant homology to the monomeric forms are considered valid potential parents in generating the monomeric Fe-hydrogenase mutants in Chlamydomonas.

While most known Fe-hydrogenases are highly O_2-sensitive, experiencing irreversible inactivation, that of Desulfovibrio vulgaris (Hildenborough) appears to be only reversibly inactivated by O_2, with the result that this sequence is highly attractive as a parent [62, 63], as is that of Pandorina morum [37]. Fe-hydrogenases in general show very highly conserved active site structures and sequences, such that the three-dimensional active site structures of the enzymes from the distantly related Clostridium and Desulfovibrio bacteria can be superimposed with a calculated deviation of only approximately 1 angstrom [23]. This feature indicates that a relatively high diversity of parental origins for the Fe-hydrogenases may result in functional progeny.

The rate at which photosynthetically-generated O_2 must be removed from an H2-producing algal bioreactor depends directly on the O_2-tolerance of the algal H2 production pathway. A thorough cost-benefit analysis of algal H2-producing bioreactor operation revealed that the viability of a commercial system depends heavily on obtaining algal mutants that can produce H2 under near-atmospheric concentrations of O2 [64]. This level of oxygen tolerance represents an approximate 70–fold improvement over wild-type Fe-hydrogenases in the concentration of oxygen tolerated (from 0.3–2 1 percent), or a 720–fold increase in the half-life of Fe-hydrogenase activity (from ~1 minute to ~1 2 hours) in the presence of atmospheric oxygen levels. In either case, it is clear that such an improvement is well within the range of improvements that have been reported for a variety of other enzymes [65–72].

2.3.5. H2 recycling. A third minimizer of photon-to-H$_2$ conversion efficiency is the presence in many green algae, including those most well-studied for H$_2$ production, Chlamydomonas reinhardtii, Scenedesmus obliquus, and Chlorella fusca, of H$_2$ uptake activity [73, 44]. However, the enzymes responsible for this activity have not yet been positively identified; in fact, the possibility remains that some of the same Fe-hydrogenases may be responsible for both H$_2$ uptake and H$_2$ production activity. The investigation of mechanisms regulating hydrogen production, involving mutants deficient in individual functions of the pathway, remains an important area of research in the development of H2 production by direct biophotolysis [11].

2.3.6. Rate and cost estimate. To date, the two-stage bioreactors making use of wild-type green algae are the only systems that have been developed for sustained production of H2 by means of direct photolysis. Rates of H$_2$ production achieved in this way are reported as 0.07–0.08 millimole per liter culture per hour [5] as seen in Table 5, the lowest rate reported for existing biohydrogen production systems. However, the substrate costs for direct photolysis are also extremely low, improving its relative efficiency in comparison with higher-rate, higher- input requiring systems.

Table 5. Comparison of rates of H2 biosynthesis [5]

BioH2 System	H2 synthesis rate (reported units)	H2 synthesis rate (converted units)
Direct photolysis	4.67 mmol H2/l/80 h	0.07 mmol H2/(l x h)
Indirect photolysis	12.6 nmol H2/μg protein/h	0.355 mmol H2/(l x h)
Photo-fermentation	4.0 ml H2/ml/h	0.16 mmol H2/(l x h)
CO-Oxidation by R. gelatinosus	0.8 mmol H2/g cdw/min	96.0 mmol H2/(l x h)
Dark-fermentations		
Mesophilic, pure straina Mesophilic, undefinedb	21.0 mmol H2/l 1/h 1,600.0 1 H2/m3/h	21.0 mmol H2/(l x h) 64.5 mmol H2/(l x h)
Mesophilic, undefined	3.0 1 H2/l/h	121.0 mmol H2/(l x h)
Thermophilic, undefined	198.0 mmol H2/l/24 h	8.2 mmol H2/(l x h)
Extreme thermophilic, pure strainc	8.4 mmol H2/l/h	8.4 mmol H2/(l x h)

[a] Clostridium species #2.

[b] A consortium of unknown microorganisms cultured from a natural substrate and selected by the bioreactor culture conditions.

[c] Caldicellulosiruptor saccharolyticus

In addition, the potential benefit of realistic research accomplishments must be taken into account. In the Thorough-cost analysis by modelers at the NREL, the cost of H$_2$ generated by direct photolysis was estimated under current laboratory conditions, as well as with improvements available in the near-term (incorporating measures definitely possible), in the long-term (incorporating reasonable research targets), and in the best case (incorporating improvements that are theoretically possible but which would require major accomplishments in several areas). The H_2 selling prices were estimated at \$5,300 per gigajoule, \$930 per gigajoule, \$110 per gigajoule, and as low as \$9 per gigajoule, respectively, for these scenarios [45].

2.3.7. Research priorities. Direct photolysis is not yet able to compete with other mechanisms of biohydrogen production in rate or cost, but the necessary advances are well within reach of modern biotechnology. To achieve practicality, the most important challenges to be overcome are the light-saturation effect and the strong repression of the Fe *hydrogenases,* both transcriptionally and post-translationally, by molecular oxygen. In addition, improvements to specific H_2 yield will be important, achievable through either modification of production hydrogenases or possibly through elimination of H_2 uptake activity. More ambitious ideas foresee advances in the light-harvesting capability of the algae, possibly by adding light-harvesting pigments to cover additional portions of the solar spectrum. The land-use issue is also extremely important, which in turn presumes that effective scale-up is achieved. While direct photolysis is a long-term prospect, its potentially low-energy requirements and the approachability of its challenges by established molecular techniques cause it to be well worth the continued research investment.

2.4. Indirect Photolysis

Indirect photolysis is the light-dependent evolution of H_2 by nitrogenases that occurs in cyanobacteria. In this process, nitrogenases evolve H_2 as they fix N_2 into NH3, simultaneously consuming ATP supplied by respiration of organic carbon fixed through photosynthesis at the rate of 16 ATP per H_2 produced [74, 5]. Although nitrogenases are highly oxygen-sensitive, like hydrogenases, they are protected to a great extent from O2, in filamentous cyanobacteria such as Anabaena and Nostoc, in specialized cells known as heterocysts. As a result, cyanobacteria undergoing indirect photolysis can evolve H2 quite rapidly and continuously under oxic conditions [12], in contrast to microbes using direct photolysis. Because N2 fixation is an inducible process that requires an absence of fixed nitrogen, cultivation in nitrate-free media is required to induce H_2 production. In addition, published H2 evolution rates are typically acquired under anoxic, light-saturated conditions that follow aerobic and frequently lower-light cultivation conditions [12].

The nitrogenase complex consists of two proteins—the dinitrogenase and the dinitrogenase reductase. The former is an $alpha_2beta_2$ heterotetramer, with alpha and beta subunits encoded by the genes nifD and nifK, respectively. The dinitrogenase reductase, encoded by nifH, is a homodimer and mediates the transfer of electrons from the external electron donor, typically a ferredoxin or flavodoxin, to the dinitrogenase [75]; mechanistic understanding of nitrogenase action is well-understood [76]. Several distinct variations of the nitrogenase structural genes exist, although they are highly conserved, and many cyanobacteria possess multiple nitrogenases. Interestingly, the unusual non-molybdenum-containing variants appear to allocate more electrons to H2 production (75). Numerous additional genes have been identified and characterized that are involved with nitrogen fixation and regulation, recently reviewed in [77].

Indirect photolysis requires a great deal of energy—16 ATP per H_2 molecule generated [34]. Furthermore, it is limited by the H_2-recycling activity of uptake hydrogenase found in all known N2-fixing cyanobacteria [13]. Another limitation is the vulnerability of filamentous cyanobacteria to fragmentation by mixing, which greatly diminishes H2 evolution capability [12]. Nevertheless, the O_2-tolerance of the process is uniquely attractive.

2.4.1. H_2 recycling. Uptake hydrogenases are found in apparently all nitrogen-fixing cyanobacteria, where they efficiently recover H_2 energy, and they thus present one of the primary limitations of indirect photolysis [75, 13]. Cyanobacterial uptake hydrogenases appear to be conserved, encoded by hupL (large subunit) and hupS (small subunit) genes; the recombinase xisC is also essential to uptake hydrogenase expression because it facilitates excision by site- specific recombination of a 10.5 kilobit element within the hupL structural gene during heterocyst differentiation [75]. In addition, bidirectional hydrogenases are present in many but not all N_2-fixing cyanobacteria. These are heterotetrameric enzymes consisting of a hydrogenase part encoded by hoxYH and a diaphorase part encoded by hoxFU, typically function in the H_2- uptake direction, and can oxidize nitrogenase-generated 112 in the absence of uptake hydrogenases.

In attempts to improve 112 evolution by indirect photolysis, several mutants have been generated in which the activity of uptake hydrogenases has been impaired [78] or in which the hupSL structural genes have been inactivated [79, 80]. These have shown significantly improved 112 production [32, 80–81], with 112 evolution rates enhanced by 3- to 10-fold over those achieved with the corresponding wild-type strains [12].

2.4.2. Cultivation conditions. Photobioreactors for hydrogen production are undergoing rapid development in attempts to provide light evenly, abundantly, and efficiently to photosynthetic cultures while facilitating gas removal (e.g., [50, 83–86]). At the same time, experimental work in indirect photolysis is investigating effects of variations in dissolved oxygen, light, temperature, nutrients and waste products, and suspension versus immobilization of cells, not to mention cell morphologies and genetic compositions [32, 75, 80]. It is not surprising, therefore, that an optimal range of conditions has not emerged, and indeed, such optimization may need to follow the identification and/or engineering of a few outstanding 112- producing organisms.

2.4.3. Rate and efficiency. Reports of photosynthetic conversion efficiency (energy in 112 produced divided by energy in photosynthetically active radiation [400–700 nm wavelengths]) vary, with the highly active strain Synechococcus sp. Miami BG04351 1 yielding a performance of 3.5 percent [82]. 112 synthesis rates achievable by indirect photolysis, however, compare quite favorably to those presently achievable by direct photolysis and by photofermentation (Table 5), at 0.355 millimoles 112 per liter culture per hour, compared to 0.07 millimoles per liter per hour for direct photolysis and 0.16 millimoles per liter per hour for photo-fermentation [5].

2.4.4. Research priorities. While the conversion efficiency of indirect photolysis is low compared to the theoretical efficiency of direct photolysis, it nevertheless presents the advantages of sustained 112 production over longer periods of time, great tolerance of aerobic conditions, and requirement for little input other than sunlight, minerals, and CO_2. In addition, several promising avenues are available for the improvement of 112 production.

The engineering of cyanobacteria to include alternative, non-molybdenum-containing nitrogenases, as well as to inactivate uptake hydrogenases, are paramount, and overexpression of nitrogenase as well as nitrogenase engineering to improve its catalytic rate have also been suggested [75]. Engineering efforts directed toward antenna improvements, minimizing light-saturation effects as well as enabling photon collection from wider regions of the solar spectrum, would also be applicable to indirect photolysis.

Among experimental organisms, the filamentous cyanobacteria have received the majority of investigation to date, but non-filamentous cyanobacteria such as Oscillatoria that protect nitrogenases from O_2 by temporal rather than physical separation of photosynthesis and N_2 fixation also deserve further examination. In addition, a great diversity of cyanobacteria adapted to a wide variety of environments exists that should be investigated for potential contributions to the 112 production effort. The ability of some cyanobacteria to grow heterotrophically is especially interesting in light of the possibility it presents to convert organic wastes into 112 [75].

Finally, further work directed toward optimizing cultivation conditions for highly desirable organisms, particularly outdoor and marine cultivation [80], is essential to the realization of commercially viable indirect photolysis.

2.5 Photofermentation

Photofermentation is the light-dependent process carried out by anoxygenic phototrophic bacteria, particularly purple non-sulfur bacteria such as Rhodobacter sphaeroides and Rhodobacter capsulatus, in which 112 is evolved by nitrogenase under nitrogen-deficient conditions, using ATP supplied by photoheterotrophic growth [19]. Although photosynthetic, this process does not split water and does not evolve O2 as in the oxygenic photosynthesis of green algae and cyanobacteria (Figure 19). It requires an anoxic atmosphere, because anoxygenic phototrophs do not protect their nitrogenases from O2 intracellularly in the way that cyanobacterial heterocysts do; however, the absence of O2 production causes the continuous production of 112 for >100h to be relatively uncomplicated, in contrast to the case with direct photolysis.

2.5.1. Cultivation conditions. Rhodobacter and other anoxygenic phototrophs are versatile organisms capable of a wide variety of growth modes, including aerobic respiration, anaerobic respiration, fermentation, and photoautotrophy, enabling them to withstand the varying culture conditions that would necessarily accompany outdoor cultivation. In addition, the nitrogenase activity of these organisms is more stable under diurnal illumination. At the same time, significant 112 production only occurs during photoheterotrophic growth, with the result that an important aspect of cultivation design will be the balance between conditions that favor nitrogenase stability and those that favor 112 production [19, 87].

Another important challenge in cultivation of all phototrophs is the self-shading effect that intensifies as cultivation volumes become larger or as cell densities become greater. This problem is already being addressed through genetic modification of the photosynthetic apparati, as it is in green algae, and a mutant of R. sphaeroides with less than half of the pigment of wild- type algae has already shown a 50 percent increase in 112 production [88].

Finally, the scale-up of photobioreactors with retention of optimal culture conditions presents several challenges of its own. Several studies have shown that 112 production is increased when cells are immobilized, and immobilization of cells on a large scale will require discovery of gels or solid supports that can minimize the problems of inhomogeneous distribution of cells and nutrients and difficulty of control, while maximizing the benefits of higher 112 production rates and a cell-free effluent. In addition, efficient mixing also poses special problems:while mixing is necessary to distribute substrates and collect gases, mechanical mixing is difficult in high surface-area-to-volume reactors, and gas sparging risks diluting the 112 produced [19].

2.5.2. Substrate range. An attractive feature of photofermentation is the high substrate-conversion efficiencies of many anoxygenic phototrophic bacteria, as well as their abilities to use a wide variety of substrates for growth and H2 production. Malate, lactate, other organic acids, sugars, and even some alcohols are used readily. The greatest potential value of photofermentation for H2 production, however, depends on the use of complex substrates such as those found in mixed organic wastes. In pursuit of this goal, numerous approaches have been attempted, and early success has been achieved with use of dairy wastewater blended with malate, sugar refinery wastewater also blended with malate, tofu wastewater, wastewater of a lactic acid fermentation plant, and olive mill wastewater. An especially promising approach for H2 production from wastewater involves the fermentative pretreatment of wastes to generate small organic acids such as lactate and malate favorable for H2 production [19].

2.5.3. Rate and efficiency. In general, H2 production by photofermentation occurs more rapidly when cells are immobilized or on a solid matrix; currently, the most rapid rates have been obtained when the cells are immobilized in porous activated glass. In addition, some substrates support more rapid H2 production than others. If an ideal laboratory system could be scaled up without diminishing the rate of H2 synthesis, rates of 3.6–4.0 liters H_2 per liter immobilized culture per hour would result, corresponding to 0.145–0.161 millimoles H_2 per liter per hour (Table 5) [5].

Photofermentation generates H2 at the expense not only of sunlight energy but also of organic substrates. Furthermore, the H2 is produced by nitrogenase, with the result that much of the input energy is diverted to N2 fixation. The result is that photofermentation has a calculated efficiency that is significantly lower than the theoretical efficiency obtainable by direct photolysis. However, great hope lies in the possibility that photofermentation may be adapted to the use of organic wastes, raising the practical efficiency a great deal [34].

In calculating efficiency, two components are typically considered: the substrate conversion efficiency and the sunlight conversion efficiency. The substrate conversion efficiency describes the percentage of a substrate utilized for H_2 production rather than biosynthesis or growth, according to the equation:

$$CxHyOz + (2x - z)H_2O - (y/2 + 2x - 2)H_2 + xCO_2$$

Although purple non-sulfur bacteria can use a wide variety of substrates for photoheterotrophic growth, only some of these are suitable for H_2 production. Substrate conversion efficiencies vary by strain; those with the highest values for the well-studied Rhodobacter sphaeroides include malate, lactate, and butyrate, ranging from 50–100 percent [19].

The light conversion efficiency, in turn, is the ratio of the total energy (heat of combustion) value of the H2 that has been obtained to the total light energy input to the photobioreactor. For photofermentative H_2 production, light energy conversion efficiencies range from 1–5 percent on average [19] substrates, is the development of cultivation techniques and/or organisms that allow the use of organic wastes. Improved understanding of the energy flow within the photofermentative H2- producing metabolism, including the mechanisms by which organic substrates improve H_2- production activity, would complement these efforts greatly and should be quite achievable through available metabolic engineering techniques. The nature of the wastes used will then inform the design of cultivation conditions, which form the next priority. Genetic modifications of antennae to limit self-shading and to allow utilization of a greater portion of the solar spectrum remain a high priority for all photobiological H_2-production techniques.

2.6. Water-gas Shift Mediated H2 Production

The water-gas shift reaction describes the oxidation of CO to CO_2 with the release of H_2:

$$CO(g) + H_2O(l) \Leftrightarrow CO_2(g) + H2(g)$$

This process is a light-independent reaction carried out by certain photoheterotrophic bacteria within the Rhodospirillaceae such as Rhodospirillum rubrum and Rubrivivax gelatinosus CBS, using an enzyme known as carbon monoxide dehydrogenase (CODH) in combination with a hydrogenase [89, 90]. Although light does not affect the CO oxidation rate, in light an uptake hydrogenase is able to oxidize the H2 to support light-dependent CO2 fixation, whereas H2 accumulates during incubation in darkness [90].

During this metabolism, the bacteria are able to use CO as the sole source of carbon and energy for ATP production. This has two important implications. First, it enables the microbes to produce additional H_2 from synthesis gas, which is a mixture composed primarily of CO and H_2 that results from the gasification of biomass. As a result, H2 production from biomass that is not readily fermented, such as lignin, becomes possible. Second, the water-gas shift process is highly thermodynamically favorable, with the result that enzymatic catalysis allows CO partial pressures to diminish rapidly and to very low levels (approaching 100 percent removal from synthesis gas). This process thus offers the possibility of a low-cost purification step for H2 before use in fuel cells, in which catalysts are rapidly poisoned by CO [5, 90].

2.6.1. Genetics. The regulation and structure of the CO dehydrogenase system of Rhodospirillum rubrum is well-characterized and is presumably similar to that of Rubrivivax gelatinosus. In R. rubrum, in the presence of CO, the CooA regulatory protein binds to the promoters of the cooFSCTJ and cooMKLXUH operons, initiating expression of the CO oxidation system. CODH, which oxidizes CO to CO_2, is encoded by the cooS gene. Electrons released by the oxidation of CO are transferred to a ferredoxin-like protein, CooF, and then through an undefined path to the CO-tolerant hydrogenase, CooH, which uses them to reduce H+ to yield H2. Functions of the other genes in the two operons have been studied as well [91, 92]. While the CODH-associated hydrogenase of R. rubrum is highly O_2-sensitive, that of R. gelatinosus is surprisingly O_2-tolerant, exhibiting a half-life near 20 hours when whole cells were stirred in air [90].

2.6.2. Rate and efficiency. H2 production by the water-gas shift reaction is one of the most rapid processes known, achieving 96.0 millimoles H2 per liter culture per hour under laboratory conditions (Table 5); estimated substrate conversion efficiencies have not yet been reported [5].

2.6.3. Research priorities. A major challenge for the water-gas shift reaction is the achievement of efficient transfer of synthesis gas into aqueous solution, as the CO must be available to the bacteria at sufficient concentrations to allow efficient metabolism [5]. Genetically, in addition, greater understanding of the interaction between CODH and the hydrogenase might indicate features amenable to optimization, and investigation of the O2-

tolerant R. gelatinosus hydrogenase to understand the basis for its resilience would be valuable to all efforts to engineer improved O2-tolerance in hydrogenases.

Further priorities, several of which the water-gas shift pathway shares with dark fermentation, are described in Section 2.7 below.

2.7. Dark Fermentation

Dark fermentation refers to the light-independent production of H_2 by anaerobic heterotrophic bacteria. While the water-gas shift process is thus technically a dark fermentation, and green algae and cyanobacteria are also capable of light-independent fermentations as described above, in the context of H2 production this term most often refers to non-phototrophic bacteria unless otherwise specified. These microbes may be mesophilic (with optimal metabolic temperatures of 25–40°C), thermophilic (40–65°C), extremely thermophilic (65–80°C) or hyperthermophilic (>80° C). They typically ferment carbohydrates (glucose, starch, or cellulose) and generate a gas of mixed composition, including not only H2 but also CO2, CH4, and/or volatile fatty acids, depending on the fermentation pathway used (Figure 17). In practice, highest H2 yields are associated with fermentations yielding a mixture of acetate and butyrate, while lower yields are associated with more reduced end-products such as propionate, lactate, and alcohols. Most fermentative bacteria are capable of multiple fermentation pathways, and culture conditions strongly influence the pathway(s) used; in particular, H2 concentrations must be kept very low to avoid end-product inhibition [93, 5].

2.7.1. Genetics. Fermentation pathways are numerous, diverse, and well-characterized genetically and physiologically in numerous organisms, particularly in gram-negative bacteria [93]; their regulatory mechanisms have been thoroughly investigated and widely-modeled (94); they have been metabolically engineered for the synthesis of numerous organic products (e.g., [95]); and numerous fermentative enzymes have been structurally characterized (e.g., [96]). As a result, the genetic, genomic, proteomic, metabolomic, and physiological groundwork has been done to facilitate a vast array of genetic and metabolic engineering efforts directed to the improvement of H2 production by dark fermentations.

2.7.2. Substrates. Since fermentations are, by definition, supported by organic substrates, an important question for dark fermentative H2 production is the availability of sufficiently inexpensive and abundant carbon sources that the resulting H2 could be commercially viable. Organic wastes are an attractive option, yet are typically complex and variable and therefore challenging to combine with sophisticated metabolic engineering approaches. In response to this problem, Logan and colleagues at the University of Pennsylvania decided to attempt 112 production with a similarly complex microbial community obtained from soil and achieved promising success [97], indicating that highly diverse, low-cost substrates may be accommodated with appropriate microbial inocula.

2.7.3. Rate and efficiency. A number of different dark fermentation systems have reported 112 synthesis rates well above 1 millimole 112 per liter culture per hour, using both pure and undefined cultures, with values reaching 121.0 millimoles 112 per liter culture per hour for undefined mesophilic cultures (Table 5). Dark fermentation, even by quite diverse systems, thus produces 112 at rates that are up to two orders of magnitude greater than those currently achieved by any of the phototrophic mechanisms.

At the same time, these high laboratory rates are achieved at the expense of purified organic substrates that would be prohibitively expensive at larger scales, especially with a substrate conversion efficiency of approximately 28 percent with readily-fermented sucrose and glucose [98, 99]. Moreover, thermophilic systems are calculated to consume unacceptably great quantities of energy to maintain high temperatures [5]. To become economically attractive, therefore, research in mesophilic dark fermentations must explore the feasibility of using organic waste streams and less-expensive organic substrates such as lignocellulose; pioneering efforts in this direction are few to date, but have already achieved early success and should be encouraged further [34].

2.7.4. Research priorities. Dark fermentation is a promising avenue for biohydrogen production that could benefit greatly from progress in a few key areas. Improvement in gas separation technology, as well as gas removal from cultures during fermentation, are uniquely crucial to economically feasible 112 production by dark fermentation, for fermentations are constrained to generate mixtures of gases and are typically subject to strong end-product inhibition [94]. Investigation of this problem is underway using hollow fiber membrane technologies, resulting in a 15 percent improvement in 112 yield, as well as using other synthetic polymer membranes, but further advances are achievable [5].

Interestingly, both of these problems may also be addressed through genetic engineering. Because most fermentation pathways are well-understood at both the genetic and physiological levels, and because many of the most attractive fermentative organisms are genetically tractable, great potential exists to enhance, diminish, or even to alter the products of specific fermentative pathways through metabolic engineering, as well as to diminish the sensitivity of crucial enzymes to end-product inhibition [34].

Further gains in fermentative 112 production will also require optimization of bioreactor designs; for example, the highest rate shown in Table 5 involved the use of activated carbon fixed-bed bioreactors that allowed retention of the 112-generating bacteria [5].

3. Research Priorities

Because of the diversity of approaches to biohydrogen production (direct and indirect photolysis, photofermentation, the water-gas shift pathway, and dark fermentation), specific research priorities have been summarized at the end of each section above. These emerging technologies have been carefully investigated for future practicability, and all face challenging problems that are nevertheless approachable by creative use of metabolic and chemical engineering. Given continued investment, each of the major biohydrogen pathways should be able to find a niche in a future sustainable-energy economy by delivering competitively-priced H2 at commercial scales.

4. Commercialization

While biohydrogen systems exist at the pilot scale that can produce H2 continuously from direct photolysis [44], indirect photolysis [12], photofermentation [19], the water-gas shift reaction [90], and dark fermentation [5], no commercial systems are yet available, and many questions regarding the practical applications of biohydrogen remain to be answered. In particular, it is not yet clear whether biohydrogen systems can be integrated with hydrogen fuel cell technologies to generate electricity at practical scales [5]. A major limitation to commercialization efforts, cited by multiple researchers, is in fact the limited communication between scientists who study biohydrogen systems and engineers who develop hydrogen fuel cell technologies. Great advancements are potentially achievable by encouraging this form of collaboration, in particular [34, 5].

REFERENCES

[1]   United States Department of Energy (2001). *Hydrogen Information Network*, http://www.eren.doe.gov/hydrogen/.

[2]   Dunn, S. (2002). *Hydrogen future: Toward a sustainable energy system*, Intl J Hydrogen Energy 27:235-264.

[3]   Dincer, I. (2002). *Technical, environmental and exergetic aspects of hydrogen energy systems*, Intl J Hydrogen Energy 27:265-285.

[4]   Elam, C., C. Padro, G. Sandrock, A. Luzzi, P. Lindblad, and E. Hagen (2003). *Realizing the hydrogen future: The International Hydrogen Agency's efforts to advance hydrogen energy technologies*, Intl J Hydrogen Energy 28:601-607.

[5]   Levin, D. B., L. Pitt, and M. Love (2004). *Biohydrogen production: Prospects and limitations to practical application*, Intl J Hydrogen Energy 29:173-185.

[6]   Zaborsky, O., Ed. (1998). *BioHydrogen.* Plenum Press, New York.

[7]   Blankenship, R. (2002). *Molecular Mechanisms of Photosynthesis.* Blackwell Science, Williston, VT.

[8]   Gottschalk, G. (1986). *Bacterial Metabolism*, 2nd Edition. Springer Verlag, New York.

[9]   Gfeller, R. P., and M. Gibbs (1984). *Fermentative metabolism of Chlamydomonas reinhardtii I: Analysis of fermentative products from starch in dark and light*, Plant Physiol 75:212-218.

[10]   Ghirardi, M. L., L. Zhang, J. W. Lee, T. Flynn, M. Seibert, E. Greenbaum, and A. Melis (2000). *Microalgae: A green source of renewable H2*, Trends Biotech 18:506-511.

[11]   Happe, T., A. Hemschemeier, M. Winkler, and A. Kaminski (2002). *Hydrogenases in green algae: Do they save the algae's life and solve our energy problems?*, Trends Plant Sci 7:246-250.

[12]   Lopes Pinto, F. A., O. Troshina, and P. Lindblad (2002). *A brief look at three decades of research on cyanobacterial hydrogen evolution*, Intl J Hydrogen Energy 27:1209-1215.

[13]   Cournac, L., G. Guedeney, G. Peltier, and P. M. Vignais (2004). *Sustained photoevolution of molecular hydrogen in a mutant of* Synechocystis sp. *strain PCC 6803 deficient in the Type I NADPH-dehydrogenase complex*, J Bacteriol 186:1737-1746.

[14]   Stal, L. J., and R. Moezelaar (1997). *Fermentation in cyanobacteria*, FEMS Microbiol Rev 21:179-211.

[15]   Troshina, O., L. Serebryakova, M. Sheremetieva, and P. Lindblad (2002). *Production of H2 by the unicellular cyanobacterium* Gloeocapsa alpicola *CALU 743 during fermentation*, Intl J Hydrogen Energy 27:1283-1289.

[16]   Garrett, R., and C. Grisham (1999). *Chapter 26. Nitrogen Acquisition and Amino Acid Metabolism, in* Biochemistry 2/e. Harcourt, Brace & Co.

[17] Whitmarsh, J., and Govindjee (2004). *The Photosynthetic Process*, University of Illinois at Urbana-Champaign, http://www.life.uiuc.edu/govindjee/paper/gov.html.

[18] Imhoff, J. F. (2004). *The Phototrophic Alpha-Proteobacteria: Nitrogen Metabolism, in* The Prokaryotes: An Evolving Electronic Resource for the Microbiological Community (M. Dworkin, Ed.). Springer-Verlag, New York,

[19] http://1 41.150.157.117: 8080/prokPUB/chaprender/jsp/showchap.j sp?chapnum=83 34.

[20] Koku, H., I. Eroglu, U. Guenduez, M. Yuecel, and L. Tuerker (2002). *Aspects of the metabolism of hydrogen production by* Rhodobacter sphaeroides, Intl J Hydrogen Energy 27: 13 15-1329.

[21] Schlegel, H. (1993). *General Microbiology*, 7th Edition. Cambridge University Press, New York.

[22] Albracht, S. P. J. (1994). *Nickel hydrogenases: in search of the active site*, Biochim Biophys Acta 1188:167-204.

[23] Adams, M. (1990). *The structure and mechanism of iron hydrogenases*, Biochim Biophys Acta 1020:115-145.

[24] Peters, J. W. (1999). *Structure and mechanism of iron-only hydrogenases: Review*, Curr Opin Struc Biol 9:670-676.

[25] Vignais, P. M., B. Billoud, and J. Meyer (2001). *Classification and phylogeny of hydrogenases*, FEMS Microbiol Rev 25:455-501.

[26] Happe, T., and J. D. Naber (1993). *Isolation, characterization, and N-terminal amino acid sequence of hydrogenase from the green alga* Chlamydomonas reinhardtii, Eur J Biochem 214:475-481.

[27] Wu, L.-F., and M. A. Mandrand (1993). *Microbial hydrogenases: Primary structure, classification, signatures, and phylogeny*, FEMS Microbiol Rev 104:243-270.

[28] Horner, D. S., P. G. Foster, and T. M. Embley (2000). *Iron hydrogenases and the evolution of anaerobic eukaryotes*, Mol Biol Evol 17:1695-1709.

[29] Schulz, R. (1996). *Hydrogenases and hydrogen production in eukaryotic organisms and cyanobacteria*, J Marine Biotechnol 4:16-22.

[30] Seibert, M., T. Flynn, D. Bensen, E. Tracy, and M. Ghirardi (1998). *Development of selection and screening procedures for rapid identification of H2-producing algal mutants with increased O2 tolerance, in* BioHydrogen (O. R. Zaborski, Ed.). Plenum Press, New York, pp 227-234.

[31] Bui, E. T., and P. J. Johnson (1996). *Identification and characterization of [Fe]-hydrogenases in the hydrogenosome of Trichomonas vaginalis*, Molec Biochem Parasitol 76:305-310.

[32] Forestier, M., S. Plummer, D. Ahmann, M. Seibert, and M. Ghirardi (2001). *Cloning of two hydrogenase genes from the green alga* Chlamydomonas reinhardtii, 2001 International Congress on Photosynthesis, Brisbane, Australia.

[33] Masukawa, H., M. Mochimaru, and H. Sakurai (2002). *Hydrogenases and photobiological hydrogen production utilizing nitrogenase system in cyanobacteria*, Intl J Hydrogen Energy 27:1471-1474.

[34] Hall, D. O., S. A. Markov, Y. Watanabe, and K. K. Rao (1995). *The potential applications of cyanobacterial photosynthesis for clean technologies*, Photosynth Res 46:159-167.

[35] Hallenbeck, P. C., and J. R. Benemann (2002). *Biological hydrogen production: fundamentals and limiting processes*, Intl J Hydrogen Energy 27:1185-1193.

[36] Greenbaum, E. (1988). *Energetic efficiency of hydrogen photoevolution by algal water splitting,* Biophys J 54:365-368.

[37] Boichenko, V. A., L. Y. Satina, and F. F. Litvin (1989). *Efficiency of hydrogen photoproduction in algae and cyanobacteria,* Fiziol Rast 36:239-247.

[38] Brand, J. J., J. N. Wright, and S. Lien (1989). *Hydrogen production by eukaryotic algae,* Biotech Bioeng 33:1482-1488.

[39] Urbig, T., R. Schulz, and H. Senger (1993). *Inactivation and reactivation of the hydrogenases of the green algae* Scenedesmus obliquus *and* Chlamydomonas reinhardtii, Z Naturforsch 48c:41-45.

[40] Boichenko, V. A., E. Greenbaum, and M. Seibert (2000). *Hydrogen production by photosynthetic microorganisms, in* Photoconversion of Solar Energy: Molecular to Global Photosynthesis (M. D. Archer and J. Barber, Eds.). Imperial College Press, London.

[41] Association of Energy Engineers (2004). *Energy Dictionary: Fuel Conversion Efficiency,* http://www.energyvortex.com/frameset.cfm?source=/ energydictionary/energyvortex.htm.

[42] Nakajima, Y., and R. Ueda (1999). *Improvement of microalgal photosynthetic productivity by reducing the content of light harvesting pigment,* J Appl Phycol 11:195-201.

[43] Nakajima, Y., M. Tsuzuki, and R. Ueda (2001). *Improved productivity by reduction of the content of light harvesting pigment in* Chlamydomonas perigranulata, J Appl Phycol 13 :95-101.

[44] Polle, J., S. Kanakagiri, J. R. Benemann, and A. Melis (2001). *Maximizing photosynthetic efficiencies and hydrogen production in microalgal cultures, in* Biohydrogen II (J. Miyake, T. Matsunaga, and A. San Pietro, Eds.). Pergamon, Elsevier, Amsterdam, pp 111-130.

[45] Melis, A. (2002). *Green alga hydrogen production: Progress, challenges, and prospects,* Intl J Hydrogen Energy 27:1217-1228.

[46] Amos, W. (2000). *Cost Analysis of Photobiological Hydrogen Production from* Chlamydomonas reinhardtii *Green Algae.* National Renewable Energy Laboratory, Golden, CO.

[47] United States Department of Energy, Office of Energy Efficiency and Renewable Energy (2004). *Solar Radiation for Energy: A Primer and Sources of Data,* http://www.eere.energy.gov/consumerinfo/factsheets/v138.html.

[48] Oak Ridge National Laboratory (2004). Bioenergy Conversion Factors, http://bioenergy.esd.ornl.gov/papers/misc/energy_conv.html.

[49] United States Department of Energy, Energy Information Administration (2004). *Home Energy Use and Costs: Residential Energy Consumption Survey,* http://www.eia.doe.gov/emeu/recs/contents.html.

[50] Turner, J. (1999). *A realizable renewable energy future,* Science 285:687-689.

[51] Gordon, J. M. (2002). *Tailoring optical systems to optimized photobioreactors,* Intl J Hydrogen Energy 27:1175-1184.

[52] Ghirardi, M. L., R. K. Togasaki, and M. Seibert (1997). *Oxygen sensitivity of algal H2- production,* Appl Biochem Biotechnol 63-65:141-151.

[53] National Renewable Energy Laboratory (1999). *Development of an Efficient Algal H2- Production System,* Golden, CO, Proceedings of the 1999 U.S DOE Hydrogen Program Review NREL/CP-570-2693 8.

[54] Heidelberg, J. F., and *et al.* (2002). *Genome sequence of the dissimilatory metal ion-reducing bacterium* Shewanella oneidensi, Nature Biotechnol 20:1118-1123.

[55] Meyer, J., and J. Gagnon (1991). *Primary structure of hydrogenase I from* Clostridium pasteurianum, Biochem 30:9697-9704.

[56] Nicolet, Y., C. Piras, P. Legrand, C. E. Hatchikian, and J. C. Fontecilla-Camps (1999). Desulfovibrio desulfuricans *iron hydrogenase: The structure shows unusual coordination to an active site Fe binuclear center*, Struct Fold Des 7:13-23.

[57] Flynn, T., M. Ghirardi, and M. Seibert (1999). *Isolation of* Chlamydomonas *mutants with improved oxygen tolerance,* American Chemical Society General Meeting, New Orleans, LA.

[58] Ghirardi, M., and M. Seibert (1999). *Process for selection of oxygen-tolerant algal mutants that produce H2 under aerobic conditions,* U.S. Patent 5,871,952.

[59] Ghirardi, M. L., T. Flynn, M. Forestier, A. Iyer, A. Melis, P. Danielson, and M. Seibert (1999). *Generation of C.* reinhardtii *mutants that photoproduce H2 from H2O in the presence of O2, in* G. Garab, Ed. Photosynthesis: Mechanisms and Effects. Kluwer Academic Publishers, The Netherlands, pp 1959-1962.

[60] Flynn, T., M. Ghirardi, and M. Seibert (2002). *Accumulation of O2-tolerant phenotypes in H2-producing strains of* Chlamydomonas reinhardtii *by sequential applications of chemical mutagenesis and selection,* Intl J Hydrogen Energy 27:1421-1430.

[61] Flynn, T., M. L. Ghirardi, and M. Seibert (2000). *Strategies for Improving Algal Hydrogen Production, in* BioHydrogen II (J. Miyake, T. Matsunaga, and A. San Pietro, Eds.). Academic Press, New York, pp 65-76.

[62] Ghirardi, M. L. (2004). *Personal communication.*

[63] van der Westen, H. M., S. G. Mayhew, and C. Veeger (1978). *Separation of hydrogenase from intact cells of Desulfovibrio vulgaris: Purification and properties,* FEBS Letters 86:122-126.

[64] Pierik, A. J., M. Hulstein, W. R. Hagen, and S. P. Albracht (1998). *A low-spin iron with CN and CO as intrinsic ligands forms the core of the active site in [Fe]-hydrogenases,* Eur J Biochem 25 8:572-578.

[65] Mann, M. K., and J. S. Ivy (1998). *Technoeconomic analysis of algal and bacterial hydrogen production systems: methodologies and issues, in* BioHydrogen (O. Zaborsky, Ed.). Plenum Press, New York, pp 415-424.

[66] Crameri, A., E. Whitehorn, E. Tate, and W. P. C. Stemmer (1996). *Improved green fluorescent protein by molecular evolution using DNA shuffling,* Nature Biotechnol 14:3 15-3 19.

[67] Crameri, A., G. Dawes, E. Rodriguez Jr., S. Silver, and W. P. C. Stemmer (1997). *Molecular evolution of an arsenate detoxification pathway by DNA shuffling,* Nature Biotechnol 15:436-438.

[68] Patten, P. A., R. J. Howard, and W. P. C. Stemmer (1997). *Applications of DNA shuffling to pharmaceuticals and vaccines,* Curr Opin Biotechnol 8:724-733.

[69] Zhang, J.-H., G. Dawes, and W. P. C. Stemmer (1997). *Directed evolution of a fucosidase from a galactosidase by DNA shuffling and screening,* Proc Natl Acad Sci USA 94:4504-4509.

[70] Chang, C., T. Chen, B. Cox, G. Dawes, W. P. C. Stemmer, J. Punnonen, and P. Patten (1999). *Evolution of a cytokine using DNA family shuffling,* Nature Biotechnol 17:793-797.

[71]   Christians, F. C., L. Scapozza, A. Crameri, G. Folkers, and W. P. C. Stemmer (1999). *Directed evolution of thymidine kinase for AZT phosphorylation using DNA family shuffling*, Nature Biotechnol 17:259-264.

[72]   Ness, J., M. Welch, L. Giver, M. Bueno, J. Cherry, T. Borchert, W. P. C. Stemmer, and J. Minshull (1999). *DNA shuffling of subgenomic sequences of subtilisin*, Nature Biotechnol 17:893-896.

[73]   Tobin, M. B., C. Gustafsson, and G. W. Huisman (2000). *Directed evolution: The "rational" basis for "irrational design" design*, Curr Opin Strucy Biol 10:42 1-427.

[74]   Gaffron, H., and J. Rubin (1942). *Fermentative and photochemical production of hydrogen in algae*, J Gen Physiol 26:219-240.

[75]   Kim, J., and D. C. Rees (1994). *Nitrogenase and biological nitrogen fixation*, Biochem 33 :389-397.

[76]   Tamagnini, P., R. Axelsson, P. Lindberg, F. Oxelfelt, R. Wuenschiers, and P. Lindblad (2002). *Hydrogenases and hydrogen metabolism of cyanobacteria*, Microbiol Molec Biol Rev 66:1-20.

[77]   Igarashi, R. Y., and L. C. Seefeldt (2003). *Nitrogen fixation: The mechanism of the Modependent nitrogenase*, Crit Rev Biochem Mol Biol 38:351-3 84.

[78]   Berman-Frank, I., P. Lundgren, and P. Falkowskia (2003). *Nitrogen fixation and photosynthetic oxygen evolution in cyanobacteria*, Res Microbiol 154:157-164.

[79]   Mikheeva, L., O. Schmitz, S. Shestakov, and H. Bothe (1995). *Mutants of the cyanobacterium Anabaena variabilis altered in hydrogenase activities*, Z Naturforsch C 50:505-5 10.

[80]   Happe, T., K. Schutz, and H. Bohme (2000). *Transcriptional and mutational analysis of the uptake hydrogenase of the filamentous cyanobacterium* Anabaena variabilis *ATCC 29413*, J Bacteriol 182:1624-1631.

[81]   Lindblad, P., K. Christensson, P. Lindberg, A. Fedorov, F. Pinto, and A. Tsygankov (2002). *Photoproduction of H2 by wildtype* Anabaena PCC 7120 *and a hydrogen uptake deficient mutant: From laboratory experiments to outdoor culture*, Intl J Hydrogen Energy 27:1271-1281.

[82]   Sveshnikov, D., K. Sveshnikov, and D. Hall (1997). *Hydrogen metabolism of mutant forms of* Anabaena variabilis *in continuous cultures and under nutritional stress*, FEMS Microbiol Lett 147:297-301.

[83]   Asada, Y., and J. Miyake (1999). *Photobiological hydrogen production*, Biosci Bioeng 88:1-6.

[84]   Markov, S. A., M. J. Bazin, and D. O. Hall (1995). *Hydrogen photoproduction and carbon dioxide uptake by immobilized* Anabaena variabilis *in a hollow-fiber photobioreactor*, Enzyme Microbiol Technol 17:306-310.

[85]   Markov, S. A., P. F. Weaver, and M. Seibert (1997). *Spiral tubular bioreactors for hydrogen production by photosynthetic microorganisms: Design and operation*, Appl Biochem Biotechnol 63-65:577-584.

[86]   Benemann, J. R. (1999). *Photobioreactors for biohydrogen production: Technology review and assessment*. National Renewable Energy Laboratory, Golden, CO.

[87]   Tsygankov, A., A. Fedorov, S. Kosourov, and K. Rao (2002). *Hydrogen production by cyanobacteria in an automated outdoor photobioreactor under aerobic conditions*, Biotechnol Bioeng 80:777-783.

[88] Koku, H., I. Eroglu, U. Guenduez, M. Yuecel, and L. Tuerker (2003). *Kinetics of biological hydrogen production by the photosynthetic bacterium* Rhodobacter sphaeroides O.U. 001, Intl J Hydrogen Energy 28:381-388.

[89] Kondo, T., M. Arakawa, T. Hirai, T. Wakayama, M. Hara, and J. Miyake (2002). *Enhancement of hydrogen production by a phtosynthetic bacterium mutant with reduced pigment,* J Biosci Bioeng 93:145-150.

[90] Bonam, D., L. Lehman, G. P. Roberts, and P. W. Ludden (1989). *Regulation of carbon monoxide dehydrogenase and hydrogenase in* Rhodospirillum rubrum: *Effects of CO and oxygen on synthesis and activity,* J Bacteriol 171:3102-3107.

[91] Maness, P.-C., and P. Weaver (2002). *Hydrogen production from a carbon-monoxide oxidation pathway in* Rubrivivax gelatinosus, Intl J Hydrogen Energy 27:1407-1411.

[92] Ensign, S. A., and P. W. Ludden (1991). *Characterization of the CO oxidation / H2 evolution system of* Rhodospirillum rubrum, J Biol Chem 266:18395-18403.

[93] Watt, R. J., and P. W. Ludden (1998). *The identification, purification, and characterization of CooJ,* J Biol Chem 273:10019-10025.

[94] Clark, D. P. (1989). *The fermentation pathways of* Escherichia coli, FEMS Microbiol Rev 5:223-234.

[95] Demain, A. L., and S. Sanchez (2002). *Metabolic regulation of fermentation processes,* Enzyme Microbiol Technol 31:895-906.

[96] Hoefnagel, M. H., Starrenburg M.J., Martens D.E., Hugenholtz J., Kleerebezem M., Van Swam I.I., Bongers R., Westerhoff H.V., Snoep J.L. (2002). *Metabolic engineering of lactic acid bacteria, the combined approach: Kinetic modelling, metabolic control and experimental analysis,* Microbiol 1 48(Pt 4): 1003-1013.

[97] Anthony, C., and P. Williams (2003). *The structure and mechanism of methanol dehydrogenase,* Biochim Biophys Acta 1647:18-23.

[98] Iyer, P., M. A. Bruns, H. Zhang, S. van Ginkel, and B. E. Logan (2004). *H2-producing bacterial communities from a heat-treated soil inoculum,* Appl Microbiol Biotechnol 66: 166-173.

[99] Kumar, N., and D. Das (2000). *Enhancement of hydrogen production by Enterobacter cloacae IIT-BT 08,* Process Biochem 35:589-593.

[100] Das, D., and T. Veziroglu (2001). *Hydrogen production by biological processes: A survey of literature,* Intl J Hydrogen Energy 26:13-28.

D. BIODESULFURIZATION OF FOSSIL FUELS

1. Introduction

Each day, more than 80 million barrels of oil are pumped from the Earth's surface. Of this vast amount, the majority (~90 percent) is processed for use in fuels, which then are combusted and released to the atmosphere in gaseous products [1]. Since sulfur constitutes up to 5 percent of crude oils, usually as organosulfur compounds, these combustion products include sulfur oxides and dioxides (SOx) that dissolve in atmospheric water vapor and ultimately yield acid rain [2, 1, 3]. In addition, SOx are also believed to reduce the efficiency of automobile catalytic converters, leading to increased tailpipe emissions of both nitrogen oxides (NOx) and CO_2.

Governments worldwide have responded by enacting laws to restrict the quantity of sulfur allowed in fossil fuels, primarily those intended for transportation. In the past 10 years, allowable levels of sulfur in transportation fuels have diminished from 2,000–5,000 parts per million (ppm) to less than 500 ppm; recent regulations proposed by the Directive of the European Parliament [4] and the EPA [5] will lower these levels to below 350 ppm. By 2010, even lower restrictions (less than 10–15 ppm in practice) are expected [1].

1.1. Hydrodesulfurization

The primary conventional technology used to remove sulfur from crude oil is hydrodesulfurization, or HDS. By subjecting crude oil to elevated temperatures and hydrogen partial pressure in the presence of a $CoMo/Al_2O_3$ or $NiMo/Al_2O_3$ catalyst, reactive sulfur components such as mercaptans, sulfides, and disulfides are converted to H_2S and hydrocarbons. Lower boiling point fractions of crude oil contain primarily these aliphatic organosulfur compounds and are therefore desulfurized with great success by HDS. In higher boiling point fractions, however, the organosulfur compounds primarily contain thiophenic rings, including thiophenes, benzothiophenes, and their alkylated derivatives. Unfortunately, HDS is much less efficient in the desulfurization of these compounds (Figure 20), with the result that so-called deep desulfurization technologies are being actively explored [7].

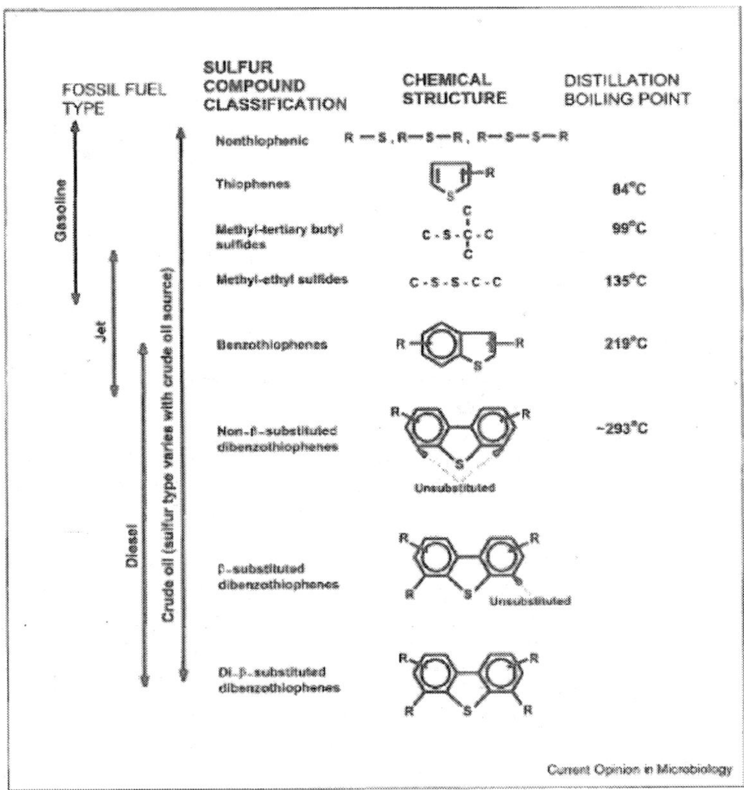

Figure 20. Organosulfur compounds present in fossil fuels [7].

2. State of the Science

2.1. Microbial Desulfurization

A promising alternative biotechnological approach employs the unusual abilities of a microbial enzymatic pathway to oxidize thiophenic sulfur atoms and subsequently cleave them from carbonaceous rings, releasing the sulfur and leaving the carbon essentially untouched [7]. This biodesulfurization (Dsz) pathway was first discovered in Rhodococcus erythropus strain IGTS8 in the late 1980s and has been studied in detail both in this organism and in analogous form in several others, including several other members of the Rhodococci as well as Agrobacterium MC501, Mycobacterium G_3, Gordona CYKS1, Sphingomonas AD109, and strains of Klebsiella, Xanthomonas, Nocardia globelula, and thermophilic Paenibacillus and Bacillus [2, 7]. Physiological studies of these organisms have been conducted primarily with the model compound dibenzothiophene (DBT) and have made great progress in establishing the platform of understanding necessary to allow further improvements through enzyme and pathway engineering [8].

2.2. Genes and Pathways

2.2.1. DBT uptake. In the native rhodococcal transformation (Figure 21), the dibenzothiophene first gains access to the cell through apparently passive means that are nevertheless assisted greatly by the tendency of rhodococci, in contrast with many other bacteria, to collect at oil-water interfaces and even to partition into the oil phase in fine emulsions [1, 9].

Despite their tendency to gather at oil drop surfaces, Rhodococci nevertheless rank low in solvent tolerance [7], in contrast to Pseudomonas strains that are typically quite solvent-tolerant [2]. To address this problem, as well as that of enzyme expression (below), researchers cloned the dszABC genes of R. erythropolis IGTS$_8$ behind the constitutive tac promoter into P. putida and P. aeruginosa species. The resulting strains grew more rapidly than the rhodococci with DBT as the sole sulfur source and converted DBT to HBP quantitatively, showing that this approach could provide strains that are more useful commercially [2].

2.2.2. Sulfur oxidation. Once within the cell, the DBT sulfur atom is first oxidized by a mono-oxygenase known as DszC. This enzyme makes use of a noncovalently-bound FMNH2, provided in reduced form by the flavin reductase DszD, to activate the molecular oxygen [8]. Notably, the availability of $FMNH_2$ is a crucial control on the rate of the desulfurization, as discussed below. A second oxidation of the sulfur atom is also catalyzed by DszC, again using FMNH2 provided by DszD, forming dibenzothiophene sulfone. A second mono-oxygenase, DszA, catalyzes the transformation of the sulfone to the sulfinate, again using FMNH2 and cleaving one carbon-sulfur bond. The remaining carbon-sulfur bond is then cleaved by a most unusual enzyme, the desulfinase DszB, releasing sulfite and hydroxybiphenyl, HBP [1, 2, 7, 8].

The availability of FMNH2 for the two mono-oxygenases appears to be a crucial rate-limiting factor in microbial desulfurization. $FMNH_2$ activity depends on the activity of DszD, the NADH-dependent FMN reductase, although DszD can be replaced in vitro by other FMN reductases [2], and FMN reduction in turn depends on the availability of NADH from cellular metabolism. This is an energy-intensive process, with ~4 NADH required per DBT desulfurized. The importance of this step has been illustrated by experiments in which flavin reductases, flavin mononucleotide reductases, or various oxidoreductases were added to the reaction mixture or over-expressed in recombinant constructs, leading to up to 100–fold increases in desulfurization rates [7].

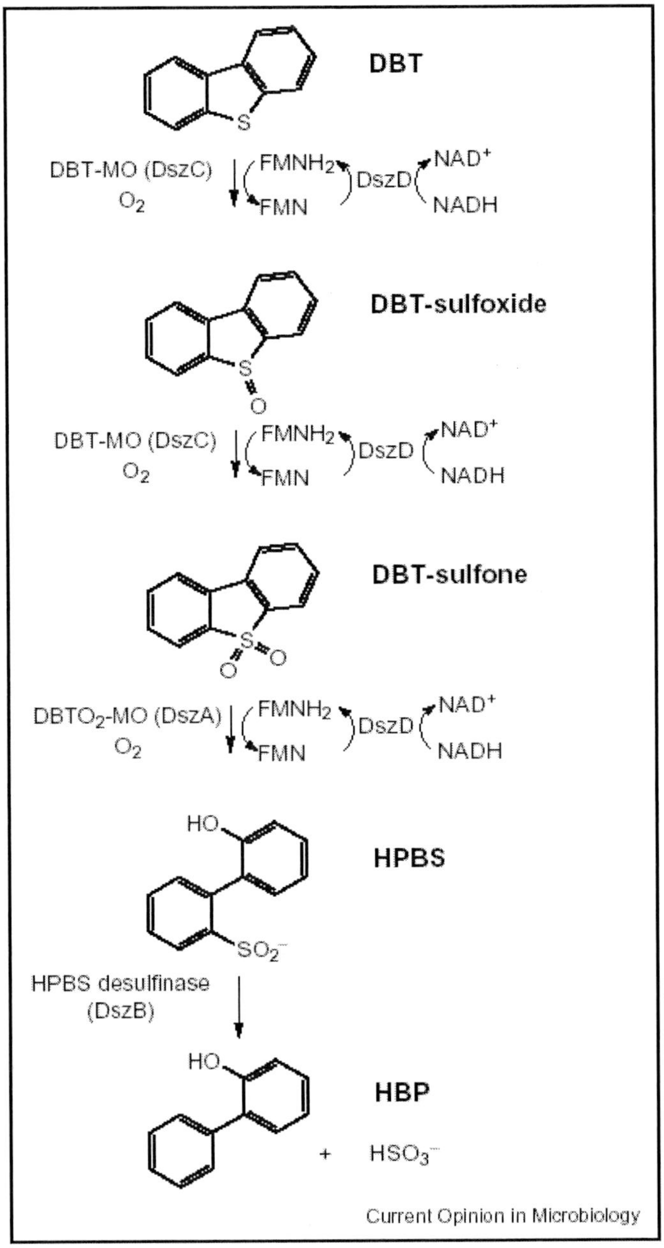

Figure 21. Dsz pathway of R. erythropus [8].

2.2.3. Substrate specificity. The enzymes of the rhodococcal Dsz pathway appear to have fairly relaxed specificities within the DBT family, desulfurizing derivatives with alkyl or aryl substitutions at rates depending on the position of the substitution: those with fewer substituents, and with substituents farther from the S atom, are generally desulfurized more slowly [10]. However, these enzymes have little activity toward the smaller thiophenes and benzothiophenes, with the result that new catalysts must be developed for efficient desulfurization of gasoline. Development of catalysts with broad specificities, especially ones that can accommodate compounds with sterically obstructed S atoms, will be essential for commercial applications [1, 7].

2.2.4. Regulation. The oxidized sulfur, released as sulfite, is used as a nutrient by the cell, and indeed the cell will not carry out desulfurization unless it is experiencing sulfur deprivation (2). The nutritional needs of the cells thus impose an additional limit on the rate of desulfurization. The HBP, all carbon atoms intact, then leaves the cell by an unknown mechanism to rejoin the nonaqueous phase (1). methionine (2) by means of a promoter responsive to sulfur-containing amino acids (1). DszD, in contrast, is encoded chromosomally. The genes for these enzymes from a variety of organisms have been cloned, sequenced, and engineered [8].

The specificity, potential to proceed nearly to completion without loss of valuable carbon, and low capital and operating costs in comparison to HDS are highly attractive features of microbial desulfurization [7], while the rate at which whole microbial cells can remove sulfur remains the greatest challenge to commercialization [8]. The throughput of substrates in this pathway may be hindered at several steps, including substrate acquisition, the supply of reducing equivalents, and enzyme turnover rates for specific substrates (Figure 22) [8].

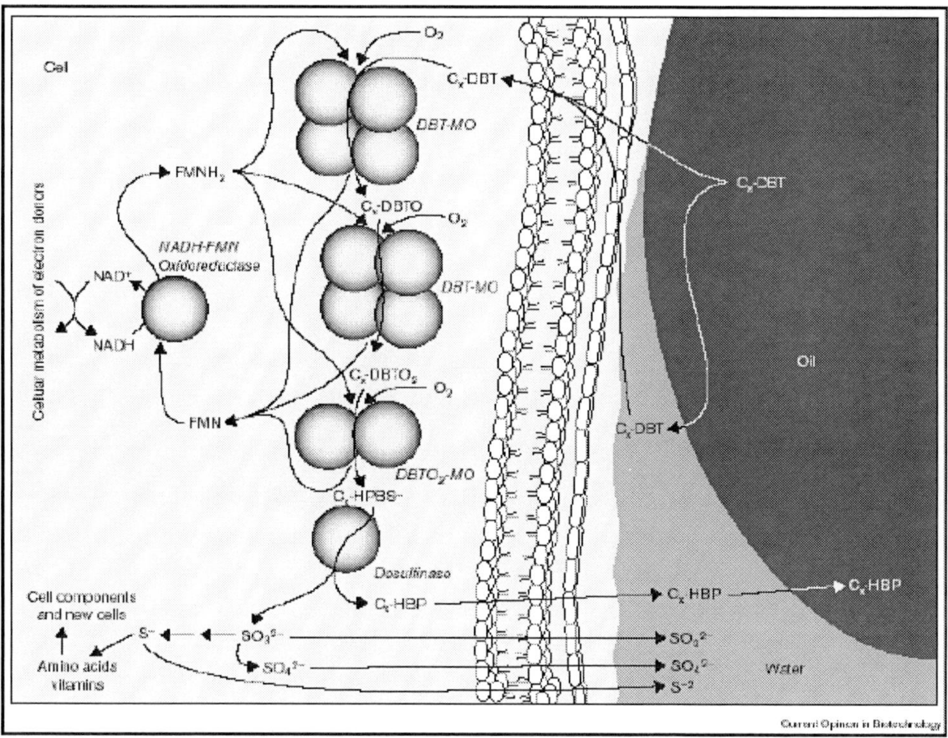

Figure 22. Conceptual diagram of biodesulfurization in R. erythropus [1].

Although the Dsz operon is transcriptionally repressed in the presence of bioavailable sulfur, the enzymes themselves (Dsz A, B, C) are not post-translationally inhibited [7], opening the possibility of overexpressing the genes under control of strong constitutive promoters. The first report of this appeared in 1997, in which researchers cloned the dszABC genes of R. erythropolis IGTS8 behind the constitutive tac promoter into P. putida and P. aeruginosa species [11]. The resulting strains grew more rapidly than the rhodococci with DBT as the sole sulfur source and converted DBT to HBP quantitatively, showing that this approach could provide strains that are quite useful commercially. The first patent on the incorporation of the Dsz genes into Pseudomonas was issued in 1999 [12], followed by another incorporating a flavin reductase (to fulfill the function of DszD) as well as the other Dsz genes into an artificial operon [13].

2.2.5. Improvements in rate and extent. Despite advances in enzyme expression, further improvements were necessary to increase the rate and extent of desulfurization to levels sufficient for commercialization; sustained rates of >20 micromoles of substrate per minute per gram catalyst were needed, far in excess of the capability of the natural Rhodococcus Dsz system [7, 1]. Between 1990 and 1998, however, optimization of biocatalyst production increased the activities of recombinant BDS catalysts 200–fold by increasing concentrations of DszA, B, and C, and optimizing conditions for DszD, bringing the catalysis rate to within an order of magnitude of that required for commercial operation [7]. Work in chemostat selection for gainof-function mutants yielded R. erythropus strains that effectively utilized octyl sulfide and 5- methyl benzothiophene [8], showing the utility of conventional approaches. At the same time, novel enzyme engineering was also required, and the combinatorial method known as RACHITT (see Chapter 2) was developed in the context of this problem. In the directed evolution of dszC genes from Rhodococcus and Nocardia, new chimeric enzymes were generated that possessed higher activity, more extensive substrate oxidation, and broader substrate specificities than either of the parents. These activities were more than sufficient to meet industrial needs and no longer limited by nutritional needs of the microbes [14, 15].

2.2.6. Tolerance to industrial conditions. Microbial desulfurization is currently most attractive as a step following the conventional HDS, which in turn requires elevated temperatures. Ideally, therefore, the biocatalytic process would require as little cooling as possible, saving energy and time; in addition, higher temperatures could afford the advantages of increased enzymatic rates and diminished contamination by other bacteria. Advances in the development of thermotolerant Dsz pathways include the discovery of desulfurization in the thermotolerant Paenibacillus, able to desulfurize DBT at 55°C [7]; the discovery of Bacillus subtilis WU-S2B, able to desulfurize DBT at 50°C (16); and Mycobacterium phlei GTIS10, also able to function at temperatures >50°C [8]. An important note in these efforts is that the thermotolerance appears not to result from the DNA sequences, which are highly similar or even nearly identical to those of the mesophilic R. erythropus IGTS8, but from other factors in the whole-cell pathway. Thermophilic catalysis may therefore be constrained to occur within a thermophilic organism.

2.3. Process Design

An important consideration in BDS process design and in choice of the microbial host is the transfer of substrate from oil into cells. Close contact between cells and oil requires emulsification of the oil-water mix that must then be broken to recover the desulfurized oil, at cost of additional time and energy. Mass transfer has been shown to limit rates of DBT degradation in Pseudomonas Dsz systems [12, 17] but presents less of a problem during use of the comparatively hydrophobic Rhodococcus cells that tend to adhere to oil-water interfaces [1]. Not only is the transfer of substrate into the cells improved, but the mixture is then also easily manipulated by means of patented devices known as hydrocyclones that readily and inexpensively separate oil-water emulsions [18]. These devices are ~1 meter long conical tubes that cause fluid to spin as it is pumped from the wide end to the narrow end, driving the denser fraction to the outside where it can be drawn off continuously. In a system where bacteria partition to the oil-water interface, the cells stay with the discontinuous phase. In a water-in-oil emulsion, cells associate with water droplets and allow separation of the clean oil phase; in an oil-in-water emulsion, cells can be concentrated with the oil for return to the reactor [1]. Additional reactor design research has reduced the influence of mass transport limitations, and current BDS reactors use staging, air sparging, and media optimization with lower water-to-oil ratios to reduce reactor size, although these conditions increase the difficulty of downstream separations [7].

As with any biological technology, the maintenance of an active catalytic population is a challenge in the context of a conventional industrial system. An important advance in this area incorporated the production and regeneration of microbial cells within the BDS process, lengthening biocatalyst activity to 200–400 hours [14] and setting an example for other bioprocesses.

3. Research Priorities

While a number of issues described above have been fully addressed, additional effort is still needed to expand the substrate specificity of the desulfurization process to include smaller compounds, allowing efficient desulfurization of gasoline, as well as to include compounds with sterically obstructed S atoms, allowing more complete desulfurization of all fossil fuels. In addition, metabolic engineering to increase the availability of $FMNH_2$ to the mono-oxygenases has the potential to improve the rate of desulfurization further, while additional increases in thermotolerance and solvent tolerance of the catalytic microbes would improve the robustness of commercial-scale systems.

4. Commercialization

Commercialization efforts in the United States were initially spearheaded by Energy BioSystems Corporation of The Woodlands, Texas, which obtained a broad patent in 1999 covering recombinant microbes based upon the Rhodococcus erythropolis $IGTS_3$ genome and worked extensively to isolate, characterize, and manipulate desulfurization genes and develop bioprocess concepts [19]. This company, later incorporated under the name Enchira Biotechnology Corporation, constructed and operated several small pilot plants, later entering an agreement with Petro Star, Inc. to design and build a 5,000 barrel-per-day BDS facility at their Valdez, Alaska refinery. The company also entered a number of technology development alliances with oil refiners, including TOTAL Raffinage Distribution S. A. of France, Koch Refining Company, Kellogg, Brown & Root, and the Exploration & Production

Division of Texaco. These efforts were discontinued in 2000, however, when Enchira discontinued development of the BDS technology [20, 21].

International government regulations are currently providing immense, annually-increasing pressure on oil refiners to diminish the sulfur content of fossil fuels below the levels attainable by current technologies. These incentives, combined with the many attractive features of biodesulfurization, are creating a highly favorable environment in which further development of biodesulfurization can occur. Nevertheless, BDS is not yet fully optimized for pilot-scale work, and basic research to optimize the rate and extent to which the Dsz pathway can remove sulfur may still require governmental support.

REFERENCES

[1]   Monticello, D. J. (2000). *Biodesulfurization and the upgrading of petroleum distillates*, Curr Opin Biotechnol 11:540-546.

[2]   Kertesz, M. A. (1999). *Riding the sulfur cycle -- metabolism of solfonates and sulfate esters in Gram-negative bacteria*, FEMS Microbiol Rev 24:135-175.

[3]   Song, C., and X. Ma (2003). *New design approaches to ultra-clean diesel fuels by deep desulfurization and deep dearomatization,* Appl Catal B: Environ 41:207-238.

[4]   European Parliament (2003). Directive of the European Parliament Relating to the Quality of Petrol and Diesel Fuels, 2003/1 7/EC.

[5]   United States Environmental Protection Agency (2000). *Control of Air Pollution from New Motor Vehicles: Tier 2 Motor Vehicle Emissions Standards and Gasoline Sulfur Control Requirements*, Federal Register 65:6698-6746, AMS-FRL-6516-2.

[6]   Babich, I. V., and J. A. Moulijn (2003). *Science and technology of novel processes for deep desulfurization of oil refinery streams: A review*, Fuel 82:607-631.

[7]   McFarland, B. L. (1999). *Biodesulfurization*, Curr Opin Microbiol 2:257-264.

[8]   Gray, K. A., G. T. Mrachko, and C. H. Squires (2003). *Biodesulfurization of fossil fuels*, Curr Opin Microbiol 6:229-235.

[9]   Kaufman, E. N., J. B. Harkins, and A. P. Borole (1998*). Comparison of batch-stirred and electrospray reactors for biodesulfurization of dibenzothiophene in crude oil and hydrocarbon feedstocks*, Appl Biochem Biotechnol 73:127-144.

[10] Folsom, B., D. Schieche, P. DiGrazia, J. Werner, and S. Palmer (1999). *Microbial desulfurization of alkylated dibenzothiophenes from a hydrodesulfurized middle distillate by Rhodococcus erythropolis I-19,* Appl Environ Microbiol 65 :4967-4972.

[11] Gallardo, M. E., A. Ferrandez, V. De Lorenzo, J. L. Garcia, and E. Diaz (1997). *Designing recombinant Pseudomonas strains to enhance biodesulfurization*, J Bacteriol 179:7156-7160.

[12] Darzins, A., L. Xi, J. Childs, D. J. Monticello, and C. H. Squires (1999*). DSZ gene expression in Pseudomonas hosts*, U.S. Patent 5952208.

[13] Squires, C. H., W. Ji, L. Xi, B. Ortego, O. Pogrebinsky, K. A. Gray, and J. Childs (1999). *Method of desulfurization of fossil fuel with flavoprotein*, U.S. Patent 5985650.

[14] Pacheco, M. A., E. A. Lange, P. T. Pienkos, L. Q. Yu, M. P. Rouse, Q. Lin, and L. K. Linguist (1999). *Recent advances in biodesulfurization of diesel fuel*, National Petrochemical and Refiners Association Annual Meeting, San Antonio, TX.

[15] Coco, W. M. (2000). *A novel method for shuffling chimeragenesis of mutated genes or of gene families from divergent bacterial genera*, 100th General Meeting of the American Society for Microbiology, Los Angeles, CA.

[16] Kirimura, K., T. Furuya, Y. Nishii, Y. Ishii, K. Kino, and S. Usami (2001). *Biodesulfurization of dibenzothiophene and its derivatives through the selective cleavage of carbon-sulfur bonds by a moderately thermophilic bacterium Bacillus subtilis WUS2B*, J Biosci Bioeng 91:262-266.

[17] Setti, L., P. Farinelli, S. Di Martino, S. Frassinetti, G. Lanzarini, and P. Pifferi (1999). *Developments in destructive and non-destructive pathways for selective desulfurizations in oil biorefining processes*, Appl Microbiol Biotechnol 52:111-117.

[18] Yu, L., T. Meyer, and B. Folsom (1998) Oil/water/biocatalyst three-phase separation process, U.S. Patent 5772901.

[19] Rambosek, J., C. Piddington, B. Kovacevich, K. Young, and S. Denome (1999). *Recombinant DNA encoding a desulfurization biocatalyst*, U.S. Patent 5,879,914.

[20] Nasser, W. (1999). Testimony of William Nasser, CEO Energy BioSystems Corporation, before the Senate Environment and Public Works Subcommittee on Clean Air, Wetlands, Private Property, and Nuclear Safety, http://epw.senate.gov/107th/nas_5-18.htm.

[21] OTC Bulletin Board (2002). ENBC - *Enchira Biotechnology Corporation*, http://www.otcbb.com/profiles/ENBC.htm.

Chapter V

SUMMARY OF FUTURE RESEARCH PRIORITIES

A. INTRODUCTION

Petroleum-based fuels and related materials are finite and expected to enter a period of diminishing availability within the next several decades. As a result, economies based on petroleum and its feedstocks will have to move towards using fuels and materials that are renewable, environmentally friendly, and of greater availability. Science and engineering communities worldwide are exploring many options. Several promising areas for future exploration and development are identified in the following sections.

B. GENETIC ENGINEERING

Current research in genetic engineering platform technologies is proceeding at an almost incomprehensibly rapid pace under the impetus of medical and basic biological research goals. While environmental biotechnology has much to gain from advances in these research goals, the support in these technologies and pace of progress are already sufficiently great, but funding for environmental goals is limited such that agencies with primarily environmental goals are encouraged to direct their support toward research priorities in other areas.

C. BIOREACTOR TECHNOLOGIES

Since bioengineering for pollution prevention involves relatively low-value products, requiring optimal bioproces sing for commercial feasibility, improvements in bioreactor technology should be a high priority in general in this field. In addition, several areas are worthy of specific mention:

- *Sensing.* Real-time sensing of gases and aqueous metabolites is of central importance because it allows or has the potential to facilitate model validation, development of descriptive kinetic expressions, and real-time process control based on sensor feedback alone and in combination with model predictions. Biosensors suitable for monitoring bioconversions in bioreactors have been previously identified as a bottleneck in the development of high-volume, low-cost processes, indicating that the development of promising emerging technologies should be encouraged in every instance possible.

- *More-detailed Structural Modeling.* In process design and optimization, the utility of a mathematical model lies in its ability to predict the operating characteristics in regions for which experimental data do not exist. Present kinetic models generally do not allow such procedures in great detail. Detailed kinetic models are also useful for capturing the dynamic responses of bioreactors to external stimuli, a set of important concerns in process control. Accordingly, a promising area of investigation is the development of more-detailed structural models that capture the salient features of complete metabolic pathways through the integration of biochemistry, molecular biology, and computational techniques. In particular, new approaches to structural kinetic modeling that include transcriptional and post-translational regulatory effects are needed.
- *Multi-scale Modeling.* In addition, new developments are needed in computational methods to capture effects of turbulence in bioreactors, effects of shearing and mechanical stresses on cellular growth and death, and the non-Newtonian nature of cellular media. These must involve multi-scale modeling to capture details at small spatial scales and must also be able to transfer relevant information to lower-resolution models that describe greater volume and time scales. CFD models are now able to establish hydrodynamic profiles among different zones of a reactor and could serve as the basis for such models, describing concentration and temperature gradients both instantaneously and over time. Such models could be extended to bridge length, volume, and time scales, linking detailed calculations at smaller scales or in critical areas with lower-resolution models that track averaged quantities. Given recent and continuing increases in inexpensive computing power, such models could contribute greatly to reactor optimization. In addition, they have the potential to guide the scale-up of industrial processes by revealing important mass and heat transfer limitations as reactor configurations are changed.
- *Applications of Genetic Engineering.* Certain challenges inherent in bioreactor operation can be greatly alleviated by the skillful application of genetic engineering technologies. For example, substrate and product-based inhibitions are common phenomena which occur when enzyme activities diminish in the presence of locally high concentrations of certain metabolites. Increasing reactor-mixing can often alleviate these problems, but diminishing the inhibitory mechanisms genetically can offer a more convenient solution. An example of this approach is presented by Agger and colleagues, who disrupted the gene responsible for glucose repression so that the glucose conversion rate did not decrease with increasing glucose concentrations that were, in turn, needed to work at high biomass concentrations. The ability of genetic engineering to solve bioreactor-based problems offers a set of great opportunities for improvements in reactor productivity.

D. Bioseparations and Bioprocessing

Separation technologies to facilitate commercial success of biomass conversions include those suitable for low-molecular-weight organic acids, organic esters, diacids, and alcohols; gases such as H2; and biobased oils such as biodiesel and biolubricants. Among these, advances in membrane technologies and in processes utilizing environmentally-benign solvents promise especially great benefits.

- *Membrane Techniques.* The development of new and/or improved membrane materials that provide increased selectivity and specificity for the desired substances, as well as increased flux with stability and robustness, is of central importance to the membrane-based techniques discussed below:

  - *Pervaporation.* The use of pervaporation to remove either water or bioproducts from bioreactor media appears promising. Continued support for new membrane materials, new module and process designs, and improved theoretical understanding and modeling of the pervaporation process should therefore be pursued. The work of Vane and colleagues at the EPA National Risk Management Research Laboratory (NRMRL) is a noteworthy example of efforts in the development of pervaporation modeling and performance prediction software.

  - *Micro- and ultrafiltration.* Microfiltration and ultrafiltration promise to become major unit operations in the emerging biorefinery arena. The development of new materials for UF and MF, including porous metals and ceramics as well as polymers, is therefore an important priority. Similarly, nanofiltration and reverse osmosis are becoming increasingly important, with recent developments in nanotechnology promising to yield new materials with significantly improved fluxes and selectivities.

  - *Membrane chromatography.* An improved understanding of the interactions between culture media components and synthetic polymers suitable for membranes would greatly facilitate the design of synthetic substrates for use in membrane chromatography. Among those, ligand-binding and sterically-interacting species should be investigated closely to improve the selectivity of membrane chromatography while maintaining acceptably high throughput.

  - *Antifouling techniques.* Fouling is a persistent problem among membrane technologies, with the result that methods to diminish fouling of membranes and ion exchange materials, as well as to remove impurities such as salts or acids that cause complications in downstream processes, are high priorities in the advancement of bioseparations.

- *Environmentally Benign Solvents.* New renewable, biodegradable solvents are needed to support environmentally-friendly extraction processes. Supercritical CO2, a highly compressed phase of CO2 possessing properties of both liquid and gas phases, is one benign solvent that has already achieved great popularity and that has the potential to contribute performance, cost-effectiveness, and sustainability to separations of both biofuels and biomaterials.

- *Integrated Modules.* Combined- or hybrid-unit operations in which a bioreactor is integrated with a bioseparation module, as in two-phase reactor systems, are particularly attractive as means to overcome limitations inherent to bioproces sing. These are particularly desirable for their potential to remove products as they are synthesized, alleviating the nearly universal problem of product inhibition in culture media.

E. POLYACTIDES

- *Development of LCIA Tools.* One of the greatest challenges facing the production of truly environmentally-benign plastic materials, PLA and otherwise, is the evaluation of net environmental impacts, beginning with feedstock production (agriculture or collection of biomass wastes), including processing steps (production of lactide and subsequent polymerization) and ending with the emissions resulting from biodegradation. Specific processes chosen at each stage, particularly concerning conventional vs. sustainable methods, are likely to have dramatic impacts on the net environmental profiles of individual materials, yet the tools with which to evaluate these differences are not yet fully refined. These metrics are needed urgently, both to guide research and development of bioplastics and to advocate use of the truly environmentally beneficial materials. Consequently, a top priority in the development of environmentally benign plastics is the continuation of efforts to develop tools and standards within the context of LCIA that will make comparisons transparent and meaningful.
- *Improvement of Physical Properties.* Presently, PLA and other biopolyesters suffer from two important deficiencies that limit their use. The first of these is their relatively low heat distortion temperatures, and the second is their relatively high permeabilities toward a number of substances, particularly water. As current, best-available LCIA analyses have indicated that PLA is indeed environmentally benign, continued research into biological, chemical, and physical transformations of PLA-based materials to improve these properties is warranted. In particular, nanocomposite technologies (Chapter III.D.) hold promise of improving both temperature distortion and permeation characteristics, as they have in conventional plastics, and should be investigated. Microcomposite technologies are related, already well-established approaches to achieve similar improvements in conventional plastics. In addition, plant microparticles derived from waste agricultural residues and simple grasses can be used directly as microparticles, providing both economic and environmental advantages. Alternatively, blending and trans-reacting PLA-based plastics with starch- or triglyceride-based materials (Chapter III.D.) may improve performance while maintaining biodegradability, with the result that these techniques also deserve further investigation. Recently, copolymerization of cellulose acetate with PLA has demonstrated that the heat distortion temperature can be increased. In this interesting case, both constituents of the plastic material come from renewable resources. This suggests that copolymerization of PLA, especially with other polymers based on renewable resources, can provide a viable route towards improved performance.

- *Exploration and Development of New Polyesters.* A recent comprehensive study by the DOE has identified 12 promising low molecular-weight materials that can be produced by fermentation in commercial quantities from plant sugars (succinic, fumaric, malic, 2,5-furandicarboxylic, 3-hydroxypropionic, aspartic, glutaric, glutamic, itonic, and levulinic acids, and the alcohols 3-hydroxybutyrolactone, glycerol, sorbitol, and xylitol). Combination of these acids and alcohols can produce polyesters by direct condensation. In particular, reactive intermediates that can be produced by anaerobic fermentations are desirable, because anaerobic processes typically lose much less of the feedstock carbon to CO2 than do aerobic processes. The success of the DuPont SoronaTM material, a polyester of such low molecular-weight precursors (1 ,4-benzenedicarboxylic acid-dimethyl ester with 1 ,3-propanediol) shows that development of sustainable processes to take advantage of readily available, renewable substances to produce additional biodegradable plastics deserves high priority for its great potential to yield both homopolymeric and copolymeric materials with new ranges and combinations of desirable properties.

F. POLYHYDROXYAKANOATES

PHA development is proceeding in a number of promising directions on both metabolic engineering and chemical engineering fronts. Fortunately, most of these have the potential for success both individually and in combination with others, such that no particular obstacle is currently forming a bottleneck to further progress. The range of physical and thermal properties achievable with PHAs is still expanding rapidly as new configurations of copolymers and blends are explored. An on-going challenge will be the ability of the metabolic engineers to keep pace with the discoveries of the materials scientists, enabling microbes to synthesize the desired polymers both conveniently and inexpensively. These efforts can be categorized as follows.

- *Investigation of Novel Polymers and Properties.* Clearly, a number of modifications of PHA composition have the potential to improve the plasticity, moldability, heat tolerance, and durability of the resulting plastics to approach those of conventional thermoplastics. Because of the promising availability and flexibility of routes to PHA synthesis, and because of increasing oil prices that will enable PHA polymers to become increasingly cost-competitive, it is a valuable effort to explore the properties of new PHA-based homopolymers, copolymers, and blends even before microbial pathways to their syntheses are in place.
- *Metabolic and Genetic Engineering.* The increasing availability of mathematical modeling tools, genomic and proteomic data and techniques, and microarray and antisense RNA technology will allow increasingly accurate prediction of useful targets for metabolic engineering. At the same time, genetic manipulation within both microorganisms and plants is becoming increasingly possible and rapid. Several enzymes central to PHA synthesis are just beginning to be explored through combinatorial and rational design mutagenesis approaches, and efforts to understand their catalytic mechanisms, substrate specificities, modes of competition with other

enzymes, and regulation are likely to contribute greatly to the microbial or plant-based syntheses of novel polymers.

- *Reactor and Processing Technology*. Gains in commercial feasibility are often found in improving bioreactor yield and in diminishing processing costs. Without reiterating topics addressed previously, it is nevertheless important to include these in the research priorities for PHA efforts, with special note that the transfer of PHA synthesis to plants may circumvent many limitations of both reactor efficiency and processing costs.

G. STARCHES, PROTEINS, PLANT OILS, AND CELLULOSICS

- *Basic Biosciences*. From a long-term perspective, continued support for basic biosciences that allow the manipulation of plants at the genetic level is absolutely essential. Only through continued development of genetic and physiological engineering techniques will the possibility of inducing organisms to produce polymers directly, and thereby reducing the cost of bioplastics by minimizing processing steps, be realized. In addition, the preference for anaerobic processes among microbial fermentations should be recognized due to the minimization of carbon loss from feedstocks. In the context of the plant-based plastics described in this section, the required tools include the genetic engineering of the cellulose-lignin and oil distribution in plants. Such basic genetic manipulation of plants is supported by the USDA. Specific strategies for pollution prevention are necessarily more narrow and should focus on developing cost-effective methods for producing plastics from plant-based matter.

- *Biodegradable Plastics*. Within the realm of starch-based plastics, commercial interest and success is currently carrying the development of compostable disposable packaging, including garbage bags, food wraps, diapers, as well as disposable food service items such as plates, cups, and utensils. Commercial research and development is even leading to improvements in water resistance and durability of starch-based materials, with the result that these are not considered to be high priorities for federal research funding. In contrast, the possibility of vastly improved strength, lightness, durability, and heat and water resistance offered by natural fiber-reinforced composites, particularly nanocomposites, cause this area to be highly attractive for additional effort. Low-cost polysaccharide-based plastic materials have the greatest potential to displace significant amounts of petroleum-based plastics.

  - *Green chemical processes*. Similarly, more benign greener chemical processes should be investigated for transforming the available renewable resources. Such activities should include enzymatic transformations when they are less energy intensive than existing chemical routes.

  - *Plant proteins*. Genetic engineering of plant proteins for specific functionality is a lower priority simply because the potential application, adhesives, is small relative to the packaging and structural uses for plastics. If plant proteins can be engineered for larger volume applications, for example, into thermoplastic film or sheet materials, the promise of coproducing both fuels and materials in an integrated biorefinery would be supported. Specialized materials produced through recombinant DNA technologies are expected to be of high commercial value but of relatively small volume.

H. BIOETHANOL

The consensus among researchers and supporters of bioethanol research, in addition to those engaged in commercial projects, is that the improvement of cellulase enzyme activity and cellulase production, both to increase the efficiency of release of fermentable sugars from biomass and to reduce cellulase cost, are two of the greatest advances needed in the effort to commercialize fuel ethanol production. In addition is the development of enzymatic pretreatment processes to release lignin from carbohydrate components and further improvement of fermentative organisms, with the particular goal of designing microbes capable of consolidated bioprocessing.

I. BIODEISEL

The cost, quality, and performance of biodiesel, as well as its overall environmental profile, could be improved by further efforts in several areas.

- *Altenative Feedstocks*. First, feedstocks other than virgin plant oils, most of which are cultivated by non-sustainable, pesticide and energy intensive agricultural practices, would ideally be explored and developed; alternatively, sustainable cultivation of oil crops should be developed. Waste oil processing technology also deserves developmental effort to allow recovery of its intrinsic energy, and microbial and algal lipid production should be investigated to determine whether they might provide feedstocks at lower cost.
- *Lipase Technology*. The further development of lipase technology will facilitate efficient enzymatic transesterifications of feedstock oils and fats and production of benign wastes with easily-recoverable coproducts, principally glycerol. Specifically, genetic engineering of lipases for greater activity and durability, as well as metabolic engineering of lipase-production pathways to understand lipase synthesis and regulation and to facilitate extracellular production, are well-positioned to offer valuable advances in enzymatic transesterification of oils and should be pursued. Research in these areas could have potentially great impacts within relatively short timescales and should be encouraged to the greatest extent possible.

J. BIOHYDROGEN

Approaches to biohydrogen production (direct and indirect photolysis, photofermentation, the water-gas shift pathway, and dark fermentation), have been carefully investigated for future practicability, and all face challenging problems that are nevertheless approachable by creative use of metabolic and chemical engineering. Given continued investment, each of the major biohydrogen pathways should be able to find a niche in a future sustainable-energy economy by delivering competitively-priced H2 at commercial scales.

- *Direct Photolysis.* Direct photolysis is not yet able to compete with other mechanisms of biohydrogen production in rate or cost, but the necessary advances are well within reach of modern biotechnology. To achieve practicality, the most important challenges to be overcome are the light-saturation effect and the strong repression of the Fe hydrogenases, both transcriptionally and post-translationally, by molecular oxygen. In addition, improvements to specific 112 yield will be important, achievable through either modification of production hydrogenases or possibly through elimination of 112 uptake activity. More ambitious ideas foresee advances in the light-harvesting capability of the algae, possibly by adding light-harvesting pigments to cover additional portions of the solar spectrum. The land-use issue is also extremely important, which in turn presumes that effective scale-up is achieved. While direct photolysis is a long-term prospect, its potentially low-energy requirements and the approachability of its challenges by established molecular techniques cause it to be well worth the continued research investment.
- *Indirect Photolysis.* While the conversion efficiency of indirect photolysis is low compared to the theoretical efficiency of direct photolysis, it nevertheless presents the advantages of sustained 112 production over longer periods of time, great tolerance of aerobic conditions, and requirement for little input other than sunlight, minerals, and CO2. In addition, several promising avenues are available for the improvement of 112 production:
  - *Engineering of cyanobacteria.* The engineering of cyanobacteria to include alternative, non-molybdenum-containing nitrogenases, as well as to inactivate uptake hydrogenases, are paramount, and overexpression of nitrogenase as well as nitrogenase engineering to improve its catalytic rate have also been suggested. Engineering efforts directed toward antenna improvements, minimizing light- saturation effects as well as enabling photon collection from wider regions of the solar *spectrum, would also be applicable to indirect photolysis.*
  - *Investigation of alternative forms of cyanobacteria.* Among experimental organisms, the filamentous cyanobacteria have received the majority of investigation to date, but non-filamentous cyanobacteria such as Oscillatoria that protect nitrogenases from O2 by temporal rather than physical separation of photosynthesis and N2 fixation also deserve further examination. In addition, a great diversity of cyanobacteria adapted to a wide variety of environments exists that should be investigated for potential contributions to the 112 production effort. The ability of some cyanobacteria to grow heterotrophically is especially interesting in light of the possibility it presents to convert organic wastes into 112.
  - *Optimizing cultivation conditions.* Finally, further work directed toward optimizing cultivation conditions for highly desirable organisms, particularly outdoor and marine cultivation, is essential to the realization of commercially viable indirect photolysis.

- *Photofermentation.* The most immediate priority for photofermentative H2 production, given its high (nitrogenase-based) energy requirement as well as its need for organic substrates, is the development of cultivation techniques and/or organisms that allow the use of organic wastes. Improved understanding of the energy flow within the photofermentative H2-producing metabolism, including the mechanisms by which organic substrates improve H2-production activity, would complement these efforts greatly and should be quite achievable through available metabolic engineering techniques. The nature of the wastes used will then inform the design of cultivation conditions, which form the next priority. Genetic modifications of antennae to limit self- shading and to allow utilization of a greater portion of the solar spectrum remain a high priority for all photobiological H2-production techniques.

- *Water-Gas Shift Mediated H2 Production.* A major challenge for the water-gas shift reaction is the achievement of efficient transfer of synthesis gas into aqueous solution, as the CO must be available to the bacteria at sufficient concentrations to allow efficient metabolism. Genetically, in addition, greater understanding of the interaction between CODH and the hydrogenase might indicate features amenable to optimization, and investigation of the O2-tolerant R. gelatinosus hydrogenase to understand the basis for its resilience would be valuable to all efforts to engineer improved O2-tolerance in hydrogenases.

- *Dark fermentation.* Dark fermentation is a promising avenue for biohydrogen production that could benefit greatly from progress in a few key areas. Improvement in gas separation technology, as well as gas removal from cultures during fermentation, are uniquely crucial to economically feasible H2 production by dark fermentation, for fermentations are constrained to generate mixtures of gases and are typically subject to strong end-product inhibition. Investigation of this problem is underway using hollow fiber membrane technologies, resulting in a 15 percent improvement in H2 yield, as well as using other synthetic polymer membranes, but further advances are achievable. Interestingly, both of these problems may also be addressed through genetic engineering. Because most fermentation pathways are well-understood at both the genetic and physiological levels, and because many of the most attractive fermentative organisms are genetically tractable, great potential exists to enhance, diminish, or even to alter the products of specific fermentative pathways through metabolic engineering, as well as to diminish the sensitivity of crucial enzymes to end-product inhibition. Further gains in fermentative H2 production will also require optimization of bioreactor designs.

K. BIODESULFURIZATION OF FOSSIL FUELS

While a number of issues described above have been fully addressed, additional effort is still needed to expand the substrate specificity of the desulfurization process to include smaller compounds, allowing efficient desulfurization of gasoline, as well as to include compounds with sterically obstructed S atoms, allowing more complete desulfurization of all fossil fuels. In addition, metabolic engineering to increase the availability of FMNH2 to the mono-oxygenases has the potential to improve the rate of desulfurization further, while additional increases in thermotolerance and solvent tolerance of the catalytic microbes would improve the robustness of commercial-scale systems.

Appendix
Contributions of the NSF/EPA Technology for a Sustainable Environment Program 1995–2004

1. Introduction

Between the years of 1995 and 2004, the NSF and EPA jointly funded the Technology for a Sustainable Environment (TSE) Program as part of the NSF/EPA Partnership for Environmental Research. This program had the goal of supporting the investigation and development of pollution avoidance and prevention processes, especially those with the potential to have long-term impact on industrial applications. Specific areas of interest included chemistry- and reaction-based engineering, non-reaction-based engineering, green design, green manufacturing, and industrial ecology for the realization of sustainable products and services. While the TSE Program was equally open to chemical, physical, mathematical, and bioengineering technologies, the Program Review in May 2004 revealed that a significant proportion of TSE grants have supported biologically-relevant endeavors. Of these, the projects that directly addressed issues raised in this State of the Science Report are summarized below.

2. TSE Contributions to Genetic and Metabolic Engineering

While the TSE program has appropriately refrained from funding projects intended primarily to advance genetic and metabolic engineering technologies, TSE-funded projects have nevertheless made excellent use of available technologies. They have also contributed substantially to the refinement of basic techniques for application in the syntheses of biomaterials and biofuels.

The following projects used metabolic engineering to develop biological pathways for the synthesis of industrially- and agriculturally-useful products, some of which are related to monomers that may be used in biomaterials. The use of biologically-generated and biodegradable starting materials to yield similarly environmentally benign products is a fundamental goal of bioengineering for pollution prevention.

- *EPA-R824726* Fermentation of Sugars to 1,2-Propanediol by Clostridium thermosaccharolyticum (PI: Cameron, University of Wisconsin-Madison)
- *EPA-R826116* Environmentally Benign Synthesis of Resorcinol from Glucose (PI: Frost, Michigan State University)
- *NSF-9819957; EPA-R826729* Metabolic Engineering of Methylotrophic Bacteria for Conversion of Methanol to Higher Value-Added Products (PI: Lidstrom, University of Washington)
- *NSF-9985421* Metabolic Engineering of Carbon Fixation and Utilization for Biopolymer Production by Cyanobacteria (PI: Stephanopoulos, MIT)
- *NSF-0118961* Metabolic Engineering of Bacillus for Enhanced Product Yield (PI: Ataai, University of Pittsburgh)
- *NSF-0124401* Metabolic Engineering of Monooxygenases for 1-Naphthol and Styrene Epoxide Formation (PI: Wood, University of Connecticut)
- *EPA-R829589* Analysis and Management of Fluxes in Bacillus Pathways for Pesticide and Protein Production (PI: Grossmann, Carnegie-Mellon University)

The engineering of microorganisms and enzymes to withstand side reactions or industrial bioreactor conditions, often facilitating more rapid and/or less expensive production, is another prominent goal of metabolic and genetic engineering that has been addressed.

- *NSF-9817621* Improving Resistance to Enzyme Alkylation During Enzyme-Catalyzed Production of Acrylamide (PI: Oriel, Michigan State University)
- *NSF-9911231*; EPA-R828562 Metabolic Engineering of Solvent Tolerance in Anaerobic Bacteria (PI: Papoutsakis, Northwestern University)

3. TSE Contributions to Bioreactor Technology, Bioseparations, and Bioprocessing

A fundamental component of any bioreactor-based sustainable technology is the ability to use non-toxic, biodegradable and/or environmentally benign solvents for the separation and processing of the desired bioproducts. The TSE Program has contributed substantially to this effort, funding projects to develop supercritical $CO2$, supercritical and near-supercritical water, polyglycols, and ionic liquids as solvents, as well as processes that avoid the need for solvents altogether.

- *NSF-96 13258* Coexisting Chemical-Biological Modifications of Chlorinated Solvents as a Basis for Waste Reduction in Pollution Prevention (PI: Watts, Washington State University)
- *NSF-9817069* Novel Compressed Solvent Extraction Processes for Enhanced Biomass Conversion by Thermophilic Bacteria (PI: Knutson, University of Kentucky)
- *EPA-R826113* Synthetic Methodology "Without Reagents" Tandem Enzymatic and Electrochemical Methods for the Manufacturing of Fine Chemicals (PI: Hudlicky, University of Florida)

- *EPA-R826117* Aqueous Processing of Biodegradable Materials from Renewable Resources (PI: McCarthy, University of Massachusetts – Lowell)
- *EPA-R828130* Nearcritical Water as a Reaction Solvent (PI: Eckert, Georgia Institute of Technology)
- *EPA-R828133* Aqueous Polyglycol Solutions as Environmentally Benign Solvents in Chemical Processing (PI: Kirwan, University of Virginia)
- *EPA-R828135* Homogeneous Catalysis in Supercritical Carbon Dioxide with Fluoroacrylate Copolymer Supported Catalysts (PI: Akgerman, Texas A&M University)
- *EPA-R828206* Development of a Heterogeneous Catalyst for Hydroformylation in Supercritical CO2 (PI: Abraham, University of Toledo)
- *EPA-R828541* Investigation of Room Temperature Ionic Liquids as Environmentally Benign Solvents for Industrial Separations (PI: Rogers, University of Alabama – Tuscaloosa)

Industrial bioprocess development also relies heavily upon the use of real-time sensors to identify promising targets for process improvements, especially in cases involving narrow profit margins. The following project addressed the development of a type of sensor that holds great promise for real-time bioreactor sensing.

- *NSF-9613556* Advanced Fluorescence-Based Environmental Sensors (PI: Thompson, University of Maryland - Baltimore County)

4. TSE Contributions to the Development of Bioplastics and Biomaterials

A pressing need within the field of biomaterials is the development of consistent, reliable tools for the assessment of the total environmental impacts of particular products. The TSE Program has supported numerous mathematical modeling efforts directed toward the assessment of product and process environmental sustainability at both small and large scales; even those not specifically intended for biomaterials applications have contributed to the general development of environmental life cycle analysis thought and are therefore included.

- *EPA-R825345* Environmentally Conscious Design and Manufacturing with Input Output Analysis and Markovian Decision Making (PI: Olson, Michigan Technological University)
- *EPA-R826114* BESS, A System for Predicting the Biodegradability of New Compounds (PI: Punch, Michigan State University)
- *EPA-R826739* New Methods for Assessment of Pollution Prevention Technologies (PI: Frey, North Carolina State University)
- *EPA-R826740* Economic Input-Output Life Cycle Assessment: A Tool to Improve Analysis of Environmental Quality and Sustainability (PI: Lave, Carnegie-Mellon University)
- *EPA-R828128* Designing for Environment: A Multi-objective Optimization Framework Under Uncertainty (PI: Diwekar, Carnegie-Mellon University)

- *EPA-R829597* Computer-Aided Hybrid Models for Environmental and Economic Life-Cycle Assessment (PI: Horvath, University of California – Berkeley)
- *EPA-R829598* Material Selection in Green Design and Environmental Cost Analysis (PI: Lin, SUNY Buffalo)
- *NSF-9985554* A Systems Ecology Approach to Life-Cycle Product Assessment and Process Design (PI: Bakshi, Ohio State University)
- *NSF-0124761* New Modeling Framework and Technology to Optimize Resource Utilization in the Plastics Supply Cycle (PI: Stuart, Purdue University)
- *EPA-R829576* Composite Resins and Adhesives from Plants (PI: Wool, University of Delaware)

The development of metabolic pathways for the synthesis of bioplastic monomers, such as 1 ,2-propanediol used in DuPont SoronaTM, of other small molecules that could become useful in biopolymeric materials, and of biopolymers themselves, are central challenges for biomaterials engineering that have been supported through several TSE grants.

- *EPA-R824726* Fermentation of Sugars to 1 ,2-Propanediol by Clostridium thermosaccharolyticum (PI: Cameron, University of Wisconsin-Madison)
- *EPA-R826116* Environmentally Benign Synthesis of Resorcinol from Glucose (PI: Frost, Michigan State University)
- *NSF-9985421* Metabolic Engineering of Carbon Fixation and Utilization for Biopolymer Production by Cyanobacteria (PI: Stephanopoulos, MIT)

In addition, the abiotic polymerization of biosynthetic monomers, as well as the biocatalysis of polymerization of other monomers, offer the potential to minimize pollution by creating biodegradable products or using non-toxic catalysts, respectively.

- *EPA-R826733* Environmentally Benign Polymeric Packaging from Renewable Resources (PI: Dorgan, Colorado School of Mines)
- *NSF-9613166; EPA-R825338* Biocatalytic Polymer Synthesis in and from Carbon Dioxide for Pollution Prevention (PI: Russell, University of Pittsburgh)
- *NSF-9728366; EPA-R826123* Development of Green Chemistry for Synthesis of Polysaccharide-Based Materials (PI: Wang, Wayne State University)
- *EPA-R828131* Biocatalytic Polyester Synthesis (PI: Russell, University of Pittsburgh)
- *EPA-831436* Plant-Derived Materials to Enhance the Performance of Polyurethane Materials (PI: Nelson, University of Massachusetts, Amherst)

Efforts to develop enzymes for non-polluting production of pulp and paper, by far the most widely-used biomaterials in developed countries, have also been supported.

- *NSF-0328031; NSF-0328033* Collaborative Research: Production and Use of a Ligninolytic Enzyme for Environmentally Benign Paper Manufacturing (PIs: Kelly, Syracuse University and Scott, SUNY College of Environmental Science and Forestry)

Finally, the sustainable production of biomass has been supported as well.

- *EPA-R831421* Cost-effective Production of Baculovirus Insecticides (PIs: Murhammer, University of Iowa; Bonning, Iowa State University)

5. TSE Contributions to the Development of Bioenergy and Biofuels

Bioethanol is one of the most promising biofuels, and the TSE Program has supported numerous projects that have addressed its primary challenges in the development of novel feedstocks, biomass pretreatment technologies, metabolic pathways to allow simultaneous fermentation of xylose and glucose resulting from lignocellulose hydrolysis, and alcohol tolerance of fermentative microbes.

- *NSF-9613342* Conversion of Paper Sludge to Ethanol and Potentially Recyclable Minerals (PI: Lynd, Dartmouth College)
- *NSF-9727096* Cellulose Conversion Using Aqueous Pretreatment and Cellulose Enzyme Mimetic (PI: Ladisch, Purdue University)
- *NSF-9817069* Novel Compressed Solvent Extraction Processes for Enhanced Biomass Conversion by Thermophilic Bacteria (PI: Knutson, University of Kentucky)
- *EPA-R826118* Development of Biotechnology to Sustain the Production of Environmentally Friendly Transportation Fuel Ethanol from Cellulosic Biomass (PI: Ho, Purdue University)
- *NSF-9911231; EPA-R828562* Metabolic Engineering of Solvent Tolerance in Anaerobic Bacteria (PI: Papoutsakis, Northwestern University)
- *NSF-9985351* Dissolution and Kinetic Fundamentals Underlying Novel Cellulosic Biomass Pretreatment Technologies (PI: Wyman, Dartmouth College)
- *EPA-R831645* Pretreatment of Agricultural Residues Using Aqueous Ammonia for Fractionation and High Yield Saccharification (PI: Lee, Auburn University)

In addition, the TSE Program has supported projects addressing both fermentative and photosynthesis-based H_2 production, as well as one investigating whole-cell biocatalysis with relevance to biodiesel production.

- *NSF-0124674* Biological Hydrogen Production as a Sustainable Green Technology for Pollution Prevention (PI: Logan, Pennsylvania State University)
- *NSF-0124821* Combinatorial Mutagenesis of a Bidirectional Hydrogenase in Chlamydomonas reinhardtii (PI: Ahmann, Colorado School of Mines)
- *NSF-0327902* Modulating Cell Permeability for Whole-Cell Biocatalysis in Chemical Synthesis (PI: Chen, Virginia Commonwealth University)
- *NSF-0328187* Functional and Structural Analysis of Algal Hydrogenase Combinatorial Mutants (PI: Ahmann, Colorado School of Mines)

INDEX

C

D

E

F

G

H

I

J

K

L

Q

R

T

U

V

W